Radio Wave Propagation of Satellite Systems

Zhenwei Zhao · Leke Lin · Changsheng Lu ·
Rui Zhang · Kun Liu · Xin Zhang

Radio Wave Propagation of Satellite Systems

国防工業出版社
National Defense Industry Press

Springer

Zhenwei Zhao
China Research Institute of Radiowave Propagation
Qingdao, China

Changsheng Lu
The Second Research Department
China Research Institute of Radiowave Propagation
Qingdao, China

Kun Liu
The Second Research Department
China Research Institute of Radiowave Propagation
Qingdao, China

Leke Lin
The Second Research Department
China Research Institute of Radiowave Propagation
Qingdao, China

Rui Zhang
The Second Research Department
China Research Institute of Radiowave Propagation
Qingdao, China

Xin Zhang
The Second Research Department
China Research Institute of Radiowave Propagation
Qingdao, China

ISBN 978-981-97-1393-6 ISBN 978-981-97-1394-3 (eBook)
https://doi.org/10.1007/978-981-97-1394-3

Jointly published with National Defense Industry Press
The print edition is not for sale in China (Mainland). Customers from China (Mainland) please order the print book from: National Defense Industry Press.

Translation from the Chinese language edition: "Wei Xing Xi Tong Dian Bo Chuan Bo" by Zhenwei Zhao et al., © National Defense Industry Press 2020. Published by National Defense Industry Press. All Rights Reserved.

This Springer imprint is published by the registered company Springer Nature Singapore Pte Ltd.
The registered company address is: 152 Beach Road, #21-01/04 Gateway East, Singapore 189721, Singapore

Contents

1 Fundamentals of Radio Wave Propagation for a Satellite System 1
1.1 Parameters of a Satellite System 2
1.1.1 Operating Frequency 2
1.1.2 Polarization and Polarization Angle 2
1.1.3 Azimuth Angle and Elevation Angle of Antennas at the Ground Station 6
1.2 Free-Space Propagation Loss (FSPL) 7
1.3 Overview of Radio Wave Propagation Effects on a Satellite System 9
1.3.1 Earth's Atmosphere 9
1.3.2 Radio Wave Propagation Effects 10
1.4 Function of Radio Wave Propagation in Designing of a Satellite System 11
References 14

2 Clear-Air Effects 15
2.1 Atmospheric Refractive Effects 16
2.1.1 Atmospheric Refractive Index [7] 16
2.1.2 Ray Bending and Effective Earth Radius [9] 17
2.1.3 Apparent Elevation Angle of a Satellite 18
2.1.4 Focusing and Defocusing Effects 20
2.1.5 Excess Radio Path Length 21
2.2 Attenuation Due to Gaseous Absorption [10] 22
2.2.1 Line-by-Line Calculation of Gaseous Attenuation 23
2.2.2 Simplified Method for Calculation of Gaseous Attenuation 31
2.3 Tropospheric Scintillation and Multipath Fading 35
2.3.1 Prediction of Scintillation Fading at Elevation Angles Greater than 5° 36
2.3.2 Prediction of Scintillation Fading for Fade Depth Greater than 25 dB 38

2.3.3 Prediction of Scintillation Fading in Transition Regions ... 39
References ... 40

3 Hydrometeor Effects ... 41
3.1 Physical Characteristics of Hydrometeors ... 41
3.1.1 Physical Characteristics of Clouds and Fog ... 41
3.1.2 Physical Characteristics of Rainfall ... 43
3.2 Cloud Attenuation ... 58
3.3 Rain Attenuation ... 61
3.3.1 Rain Attenuation Rate ... 61
3.3.2 ITU-R Rain Attenuation Prediction Method [27] ... 63
3.3.3 Prediction Methods in China ... 66
3.3.4 Worst-Month Rain Attenuation ... 68
3.4 Rain Attenuation Frequency Conversion ... 70
3.4.1 Instantaneous Frequency Conversion ... 71
3.4.2 Long-Term Statistical Frequency Conversion of Rain Attenuation ... 74
3.5 Dynamic Characteristics of Rain Attenuation ... 74
3.5.1 Fading Duration ... 75
3.5.2 Slope of Fading ... 77
3.6 Rain Attenuation Time Series ... 79
3.7 Short-Term Forecasting of Rain Attenuation ... 81
3.8 Rain-Induced Cross-Polarization ... 83
3.8.1 Description of Depolarization ... 83
3.8.2 Rain-Induced Depolarization Mechanism ... 84
3.8.3 Prediction Method of Rain-Induced Cross-Polarization ... 85
3.8.4 Frequency Scaling of Rain-Induced Cross-Polarization ... 87
3.9 Other Propagation Effects ... 88
3.9.1 Wet Antenna Attenuation ... 88
3.9.2 Attenuation Due to Sand and Dust ... 88
3.10 Total Atmospheric Attenuation ... 89
3.11 Radio Wave Environment Data in Prediction Model ... 89
3.11.1 ITU-R Digital Maps of Radio Environment ... 90
3.11.2 China's Digital Map of Radio Environment ... 90
3.11.3 Selection Principle of Radio Environment Parameters ... 93
References ... 94

4 Earth's Surface Reflection and Local Environment Effects ... 97
4.1 Overview ... 97
4.2 Radio Wave Propagation Effects of Maritime Mobile Satellite Service ... 99

4.2.1 Fading Due to Sea Surface Reflection ... 100
4.2.2 Prediction of Interference from Adjacent Satellites ... 103
4.3 Radio Wave Propagation Effects of Land Mobile Satellite Service ... 105
4.3.1 Multipath Model for Line-of-Sight Conditions ... 106
4.3.2 Shadowing Effects ... 108
4.4 Radio Wave Propagation Effect of Aeronautical Mobile Satellite Service ... 112
4.4.1 Fading Due to Sea Surface Reflection ... 112
4.4.2 Multipath Model for Aircraft During Approach and Landing ... 114
4.5 Analysis of Propagation Effect for Non-GEO Satellite Service ... 120
References ... 123

5 Ionospheric Effects ... 125
5.1 Ionospheric Characteristics ... 125
5.1.1 Overview of the Ionosphere ... 125
5.1.2 Ionospheric Models ... 127
5.1.3 Ionospheric Total Electron Content (TEC) ... 130
5.1.4 Ionospheric Irregularities ... 133
5.2 Background Ionospheric Propagation Effects ... 135
5.2.1 Ionospheric Refractive Index ... 136
5.2.2 Group Delay ... 137
5.2.3 Phase Advance ... 138
5.2.4 Refraction ... 140
5.2.5 Dispersion ... 142
5.2.6 Faraday Rotation ... 145
5.2.7 Doppler Effect ... 146
5.2.8 Absorption ... 147
5.3 Ionospheric Scintillation ... 149
5.3.1 Ionospheric Scintillation Characteristics ... 149
5.3.2 Scintillation Amplitude Distribution and Spectrum Characteristics ... 153
5.3.3 Statistics Characteristics of Ionospheric Scintillation ... 154
5.3.4 Ionospheric Scintillation Characteristics in China Region ... 161
References ... 166

6 Radio Noise and Interference Coordination ... 167
6.1 Definition and Classification of Radio Noise Signals ... 167
6.2 Solar and Cosmic Noise ... 168
6.2.1 Galactic Noise ... 169
6.2.2 Noise Radiation from Celestial Bodies ... 169

6.3 Noise from Earth's Surface and Atmospheric Noise 171
6.3.1 Uplink Noise 171
6.3.2 Downlink Noise 173
6.4 Coordination Area of Earth Station 177
6.4.1 Coordination Distance 178
6.4.2 Coordination Distance Considering Propagation in Clear Air 180
6.4.3 Coordination Distance Considering Hydrometeor Scattering 187
References 194

7 Fade Mitigation Techniques and Propagation Analysis of Satellite System Design 195
7.1 Link Performance Analysis of Satellite Communication Systems 196
7.1.1 Carrier-to-Noise (C/N) Ratio 196
7.1.2 Bit Error Ratio 199
7.2 Fade Mitigation Techniques for Satellite Communication Systems 201
7.2.1 Diversity 201
7.2.2 Power Control 206
7.2.3 Adaptive Signal Processing 208
7.3 Propagation Analysis of Typical Satellite System Design 209
7.3.1 Propagation Analysis of Fixed Satellite System Design 210
7.3.2 Propagation Analysis of Mobile-Satellite Service System Design 216
References 219

Chapter 1
Fundamentals of Radio Wave Propagation for a Satellite System

Reliable information about the propagation characteristics serves as the backbone for the design, fabrication, and operation of a satellite system. The transmission of radio signals from the ground users to the satellite in such a system must pass through the earth's atmosphere, including the troposphere and the ionosphere layers. The effect generated on radio wave propagation by the earth's atmosphere, the ground, and different coverings brings about a change in amplitude, phase, polarization, and angle of arrival of the Earth-space link signal, resulting in signal degradation and even grave system interruption; hence, it is vital to consider the effect of radio wave propagation on system performance as well as achievement of projected technical indexes in the system design [1–3]. How radio wave propagation affects the functioning of the satellite system depends on the operating frequency, transmission method, angle of elevation toward the satellite, and local geographic, meteorological, and ionospheric environment. For example, while rain attenuation plays an important role in satellite systems operating at the Ku frequency band and above, it rarely affects satellite systems operating at the L frequency band; similarly, clutter reflections, fading, and shadowing emerge as instrumental factors responsible for affecting the performance of a mobile satellite system and are preventable by the selection of a proper terminal site in the fixed satellite system. While planning and designing a satellite system, a reliable formulation of the transmission link must be carried out in accordance with the statistical characteristics of radio wave propagation for the Earth-space link pertaining to diverse satellite systems, to enhance the environmental suitability of the link. During the functioning of a satellite system, fade dynamics, short-term forecasting, and near real-time forecasting can be utilized to ensure the smooth implementation of anti-fading techniques as well as the planning of system tasks.

This chapter primarily introduces some fundamental knowledge about satellite systems and radio wave propagation.

Z. Zhao et al., *Radio Wave Propagation of Satellite Systems*,
https://doi.org/10.1007/978-981-97-1394-3_1

1.1 Parameters of a Satellite System

1.1.1 Operating Frequency

The operating frequency of a satellite system must be greater than 30 MHz, since radio waves lower than 30 MHz are generally reflected by the ionosphere while passing through the earth's atmosphere. The operating frequency of a satellite system is decided by its service type. The satellite communication system always uses spectrum window bands present between atmospheric absorption peaks to reduce the absorption loss and boost the transmission rate by opting for higher and higher frequencies. Nonetheless, regarding earth resources satellites, their operating frequency depends on their purpose of detection and remote sensing. For example, the infrared band is utilized for remote sensing and detection of the temperature of the earth's surface, and plant and cloud cover, the microwave band for earth's surface and ice sheet thickness detection, the millimeter band for cloud and mist detection, water vapor absorption band for remote sensing of water vapor, and oxygen absorption band for remote sensing of dry air.

With the expansion of satellite services and development of relevant techniques, the operating frequency of the satellite systems has improved. Satellite systems operating at VHF, UHF, L, and C bands are widely employed in fields such as communication, remote sensing, and navigation; and satellite systems operating at Ku- and Ka-bands also appear prevalent among numerous monitoring and detection fields, while satellite systems operating at EHF band and above are gradually being developed and implemented in various similar fields (Table 1.1).

By and large, there are two methods to characterize the bands. One of them is as shown in Fig. 1.1, which is based on the application of the radar system in 1940s, but the boundary division of the frequency bands for the actual application is not very rigid. The K band is further divided into Ku- and Ka-bands, where the frequency overlap occurs quite frequently; for example, while the Ku-band covers the frequency range from 10.9 to 12.5 GHz in some articles, it ranges from 11.7 to 14.5 GHz in others. The other one is the band division method used by the International Telecommunication Union (ITU). As shown in Table 1.2, this method divides the wavelength ten-fold, making the data much more explicit than the first method. According to this method, the frequencies utilized for the chief services provided by a satellite range centrally between VHF and EHF bands; in particular, the majority of the contemporary satellite systems operate at the SHF band, which is a challenging frequency range to characterize meticulously.

1.1.2 Polarization and Polarization Angle

Although radio waves radiate outwards spherically, they are perpetually deemed to be plane waves when moving away from the emission source, and the electrical

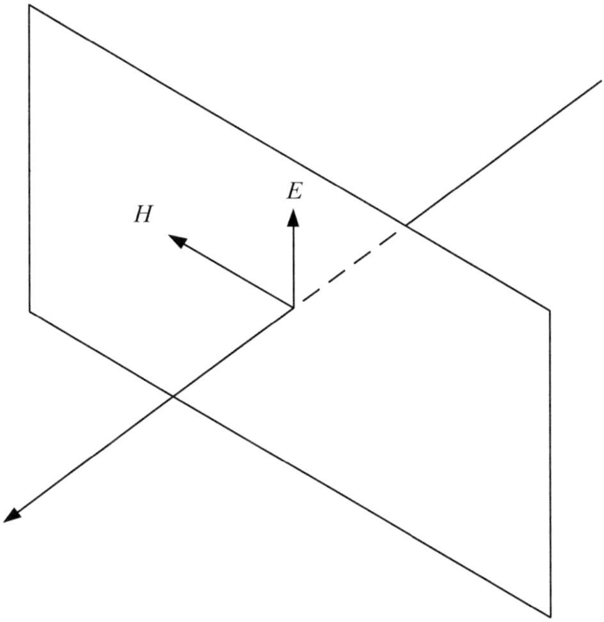

Fig. 1.1 Plane wave and polarization

Table 1.1 Frequency band division characterized by alphabets [3]

Symbol	Radar		Space communication	
	Spectrum area/GHz	Frequency band/ GHz	Nominal frequency/ GHz	Frequency band/ GHz
L	1–2	1.215–1.4	1.5	1.525–1.71
s	2–4	2.3–2.5 2.7–3.4	2.5	2.5–2.690
C	4–8	5.25–5.85	4/6	3.4–4.2 4.5–4.8 5.85–7.075
X	8–12	8.5–10.5		
Ku	12–18	13.4–14.0 15.3–17.3	11/14 12/14	10.7–13.25 14.0–14.5
K	18–27	24.05–24.25	20	17.7–20.2
Ka	27–40	33.4–36.0	30	27.5–30.0
V	–	–	40	37.5–42.5 47.2–50.2

Table 1.2 Common band division by ITU [3]

Frequency band	Frequency range	Wavelength
ULF	300 Hz–3 kHz	10^5–10^6 m
VLF	3 kHz–30 kHz	10^4–10^5 m
LF	30 kHz–300 kHz	10^3–10^4 m
MF	300 kHz–3 MHz	10^2–10^3 m
HF	3 MHz–30 MHz	10–100 m
VHF	30 MHz–300 MHz	1–10 m
UHF	300 MHz–3 GHz	10 cm–1 m
SHF	3 GHz–30 GHz	1 cm–10 cm
EHF	30 GHz–300 GHz	1 mm–1 cm

field is vertical to the magnetic-field vector, located in the plane perpendicular to the direction of propagation at that moment, as shown in Fig. 1.1 [4].

Considering that radio waves propagate in space, if the direction of the electrical-field vector does not change, the waves undergo linear polarization at that moment, and if the electrical-field vector rotates around its propagation axis, the waves are circularly polarized. The circular polarization entails two orthogonal linearly polarized electric vectors of equal amplitude with a 90° phase difference. Elliptical polarization forms the most general type of polarization for electromagnetic radiation, with linear and circular polarization being its special cases. The polarization method adopted by a system depends on various factors. As a general principle, the negative effect of propagation on linearly polarized waves must be smaller than that on circularly polarized waves. At the same time, the circularly polarized antennas deliver far more benefits than the linearly polarized antennas, such as simple structure, ready manufacturability, and low cost. However, when radio waves exhibit circular or rotatory orientation owing to the presence of media, the circularly polarized waves are deemed more suitable since the low-frequency radio waves usually pass through the ionosphere.

The polarization angle is applied to establish whether the direction of the linearly polarized waves emitted by the satellite overlaps with that of the receiving antennas at the ground station. According to the electromagnetic field theory, plane waves with the electric field intensity perpendicular to the incident plane are defined as perpendicularly polarized waves, and plane waves with the electric field vector parallel to the incident plane are defined as parallelly polarized waves. For engineering application, plane waves with the electric field vector parallel to the ground surface are defined as horizontally polarized plane waves, and plane waves with the electric field intensity perpendicular to the ground surface are defined as vertically polarized plane waves, and these definitions with reference to earth's surface appear more direct in nature. With respect to the linearly polarized waves emitted by the satellite, the horizontal polarization is generally defined as electric field vector being parallel with the equatorial plane, while the perpendicular polarization is defined as electric field

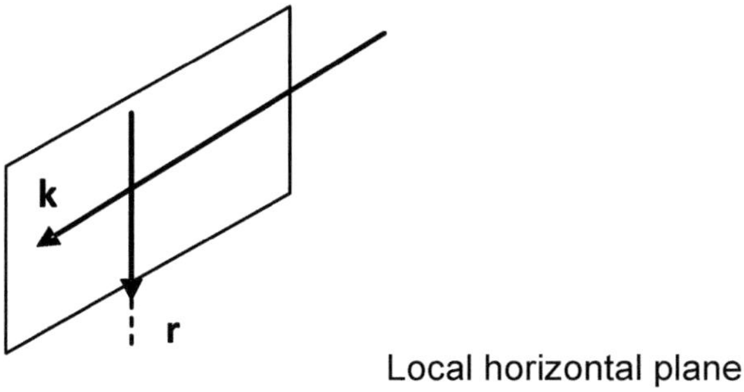

Fig. 1.2 Sketch of reference surface

vector being parallel with the polar axis of the earth. In this case, these two polarizations act parallel to the ground surface at the sub-satellite point of the equator, and the polarization vector forms a certain angle with reference to the earth's surface at other points in the beam cover areas of the satellite. According to preceding research findings, the plane in which the propagation direction is perpendicular to the local gravity direction (plumb line) should be selected as the reference surface, as shown in Fig. 1.2. k in the figure refers to the propagation direction of electromagnetic waves, and r denotes the direction of local gravity.

If observing from the ground station toward the satellite, the turner should be adjusted according to the following angles with respect to the original polarization direction

$$\psi = \arccos \frac{\sin\phi_L\{1 - [R_E/(R_E + R_0)]\cos L \cos\phi_L\}}{\left\langle(\cos^2\phi_L \sin^2 L + \sin^2\phi_L)\{[R_E/(R_E + R_0)]^2\cos^2\phi_L + 1 - 2[R_E/(R_E + R_0)]\cos L\cos\phi_L\}\right\rangle^{1/2}} \tag{1.1}$$

In the formula: R_E refers to earth's radius (km), R_0 refers to height of a geostationary satellite above the earth (km), ϕ_L refers to absolute value of latitude of the ground station (°).

L is calculated using the following formula:

$$L = |\theta_S - \theta_L| \tag{1.2}$$

In the formula: θ_s refers to longitude of sub-satellite point (°); ϕ_L refers to longitude of the ground station (°).

Because $R_0 >> R_E$, formula (1.1) can be simplified as

$$\psi = \arctan \frac{\sin L}{\tan\phi_L} \tag{1.3}$$

In the northern hemisphere: If the ground station is located at west of the meridian of the satellite, ψ refers to the angle that should be rotated counterclockwise; if the

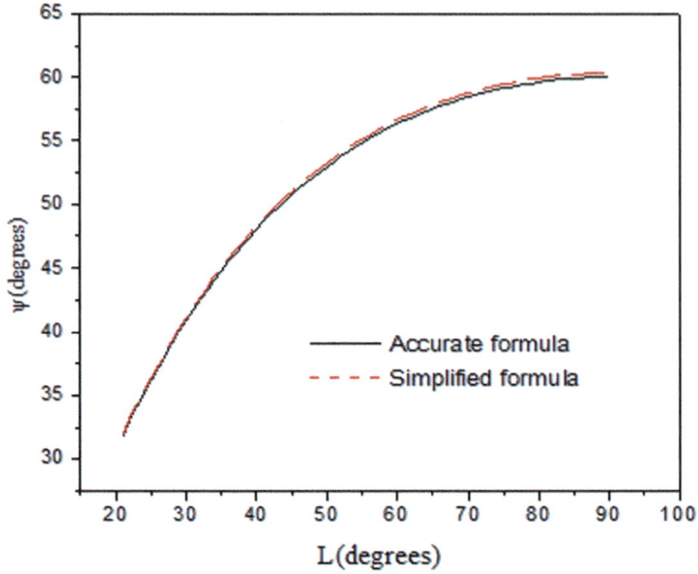

Fig. 1.3 Comparison between accurate formula (1.1) and simplified formula (1.3)

ground station is located at east of the meridian, then ψ refers to the angle that should be rotated clockwise. The conditions are reversed for the southern hemisphere. The final polarization angle of the receiving antennas is the sum of (counterclockwise) or the difference between (clockwise) its initial polarization angle (0° for horizontal polarization and 90° for perpendicular polarization) and ψ.

The calculation result for formula (1.3) is almost identical with that of formula (1.1), Fig. 1.3 provides the result for formula (1.2) and formula (1.3) at 30° of northern latitude changing with L, with max error not more than 0.3° [5].

1.1.3 Azimuth Angle and Elevation Angle of Antennas at the Ground Station

For radio wave propagation prediction and satellite system application, identifying the elevation angle and azimuth angle of antennas at the ground station is essential, as shown in Fig. 1.4. The following methods can be used to calculate these angles.

(a) Azimuth angle;
(b) Elevation angle.

Firstly, we calculate a middle angle A':

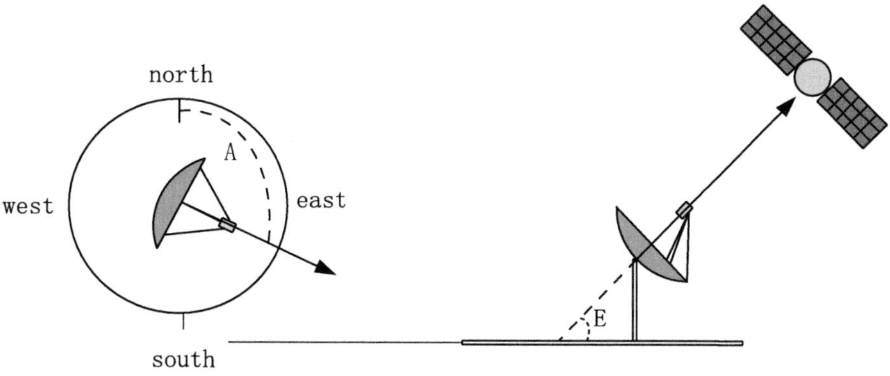

Fig. 1.4 Azimuth angle and elevation angle of antennas at the ground station

$$A' = \tan^{-1}\left(\frac{\tan|\theta_S - \theta_L|}{\sin\phi_L}\right) \tag{1.4}$$

In the formula: θ_L refers to longitude of the ground station (°); Φ_L refers to latitude of the ground station (°); and θ_S refers to longitude of the sub-satellite point (°).

The formula of azimuth angle A is shown as follows because the relative position of the ground station and the sub-satellite point is different.

When the ground station is located at the northern hemisphere and the satellite is at south by east, $A = 180° - A'$; and when the satellite is at south by west, $A = 180° + A'$. When the ground station is located at the southern hemisphere and the satellite is at north by east, $A = A'$; and when the satellite is at north by west, $A = 360° - A'$.

The formula of the elevation angle A is

$$E = \tan^{-1}\left\{\left[\cos\phi_L\cos(\theta_L - \theta_s) - \frac{R_E}{r}\right] \Big/ \sqrt{1 - \left[\cos\phi_L\cos(\theta_L - \theta_s)^2\right]}\right\} \tag{1.5}$$

In the formula: r refers to the orbital radius of the geostationary satellite, $r =$ 42164.2 km; R_E refers to the earth's radius, $R_E = 6378.155$ km.

1.2 Free-Space Propagation Loss (FSPL)

The free space denotes the propagation space devoid of any medium attenuation, obstacle, and multipath effects. This space is an ideal concept that only exists hypothetically. It is normally acknowledged that as long as the atmosphere above ground comprises the isotropic homogenous media, its relative dielectric constant ε, and relative magnetic conductivity μ are both equal to 1, there is no obstacle in the propagation path, and the field intensity of the reflected signal toward the receiving antennas stationed at the ground can be ignored. Thereby, this entire scenario can be considered as free-space propagation. Free-space radio wave propagation, regarded as the most elementary and simplest propagation method for radio waves, provides an effortless propagation environment for radio wave propagation. The free-space propagation loss, which is caused by the diffusion of radio waves radiated from antennas into space, is the basic propagation loss for radio propagation of a satellite system, and while there are normally no other effects attenuating the signal in free space situation, other propagations could produce added effects with respect to the free-space propagation loss.

When radio waves radiating from a point source propagate in free space, their energy is evenly diffused into a spherical shape. In case of directive antenna, radio waves are centered in one direction, and they gain in this direction. If transmission power of transmitter is P_t, gain is G_t, effective receiving area of receiving antennas is A_r, path length is d, media loss of path is L, the receiving power of receiving antennas is as shown in the following formula as per diffusion rule of signals in free space.

$$P_r = \frac{P_t G_t}{4\pi d^2} \cdot \frac{A_r}{L} \tag{1.6}$$

In this formula, the effective receiving area of receiving antennas A_r can be expressed by the gain of receiving antennas G_r and wavelength λ, that is

$$A_r = G_r/(4\pi/\lambda^2) \tag{1.7}$$

We can reduce the formula as shown below by plugging formula (1.7) into formula (1.6):

$$P_r = P_t G_t G_r/(4\pi d/\lambda)^2 = P_t G_t G_r/L_0 \tag{1.8}$$

In the formula: $L_0 = (4\pi d/\lambda)$ denotes the free-space propagation loss. This value is in dB and after converting wavelength into frequency, we can obtain the final formula as

$$[L_0] = 92.4 + 20\lg(d) + 20\lg(f) \quad \text{dB} \tag{1.9}$$

In the formula: Distance d is expressed in km, frequency f in GHz.

For example, assume the distance between the ground station and the fixed satellite to be 37,000 km, operating frequency is 20 GHz, then the calculation using formula (1.9) delivers 209.8 dB of free-space attenuation. This value seems very large, signifying that the free-space attenuation in the satellite system under consideration forms the main transmission loss.

According to formula (1.9), the free-space attenuation intensifies with a rise in frequency, and the relationship between gain and frequency of the antennas is as shown below:

$$G = 10\lg\left(\frac{4\pi}{\lambda^2}A_{eff}\right) = 10\lg\left(\frac{4\pi}{c^2}f^2A_{eff}\right) = 21.45 + 20\lg f + 10\lg A_{\text{eff}} \ \text{dB} \tag{1.10}$$

In the formula: A_{eff} is the effective area of the antennas

According to formula (1.10), when the effective area of the antennas remains constant, the gain of antennas increases with an increase in frequency. If the effective area of the antennas does not change, the total receiving power can be calculated using the formula (1.8):

$$\begin{aligned}[P_r] &= [P_t] + [G_t] + [G_r] - [L_0] \\ &= [P_t] - 49.5 + 20\lg f - 20\lg d + 10\lg A_t + 10\lg A_r\end{aligned} \tag{1.11}$$

In the formula: A_T, A_R are effective areas of transmission and receiving antennas, respectively

1.3 Overview of Radio Wave Propagation Effects on a Satellite System

Although the free-space propagation loss is regarded as the most pivotal loss found in satellite-ground links, its value is a fixed value with respect to fixed links and can be easily predicted while planning the system design since it holds a fixed relationship with distance. However, when the radio waves propagate between the satellite and the ground, the high-altitude ionosphere, low-altitude atmosphere, and multiple intricate features of the earth's surface come into play and can generate arbitrary complex propagation effects on the radio waves. The reliability of the links of a satellite system could be highly compromised if these effects are overlooked while designing the system, and thus, these effects must be viewed as critical factors that are capable of greatly influencing the environmental suitability of satellite links in a satellite system design.

1.3.1 Earth's Atmosphere

Under the influence of the earth's gravitational force, various components of the atmosphere surrounding our planet get engaged to form the earth's atmosphere. Owing to the earth's rotation and difference in the solar radiation absorption degree of atmosphere layers at different altitudes, the atmosphere is even in the horizontal direction and shows an evident distribution of layers in the perpendicular direction, thus, the atmosphere can be divided into multiple layers in accordance with different thermodynamic properties, ionospheric conditions, and atmosphere components. By and large, there are several methods to define the atmospheric strata, including ① being divided into troposphere, stratosphere, mesosphere, and thermosphere by thermal structure of neutral components; ② being partitioned into homosphere and heterosphere by chemical components of the atmosphere; ③ being divided into neutrosphere, ionosphere, and magnetosphere by electromagnetic properties of the atmosphere; ④ by pressure structure of the atmosphere—the atmosphere from 500 to 2,000–3,000 km above ground level is called as exoatmosphere or exosphere that gradually transits to the interplanetary space, and the space below interplanetary space is called as barosphere [6, 7]. In radio wave propagation research, we divide the atmosphere into the troposphere, stratosphere, ionosphere, and magnetosphere, as shown in Fig. 1.5 [8, 9]. During the propagation of radio waves at common frequency bands, the waves are affected the most in the troposphere and ionosphere. The atmospheric environment influencing radio wave propagation is also called as spatial radio environment, in which, the troposphere, as the atmospheric layer closest to the ground, exhibits evident convective motion. In general, the highest altitude of the troposphere around the equator and tropic ranges from 15 to 20 km, that of the polar region and mid-latitude zone ranges from 8 to 14 km, and hence, the 3 GHz frequency band and above (SHF, EHF, THF) gets mainly affected by the troposphere layer. The ionosphere, being the ionized space of earth's atmosphere, usually refers to the space ranging from 60 to 1,000 km above the ground level, and the 3 GHz frequency band and below (SHF, UHF, VHF) mainly gets affected by it.

1.3.2 Radio Wave Propagation Effects

The radio wave signals of a satellite system pass through the entire atmospheric envelope. The atmospheric components of the troposphere, layering structure, atmospheric turbulence, and meteorological phenomena such as cloud, snow, rain, fog, ionospheric structure, space changes, and irregularities produce diverse propagation effects and influences on radio waves; and the ground (or sea) surface reflection, terrain, and surface features scattering, diffraction, and shadowing tend to affect a satellite signal very differently in terms of propagation as the signal approaches the earth's surface. The chief propagation effects influencing the performance of a satellite system are provided in Table 1.3 and described at length in subsequent chapters.

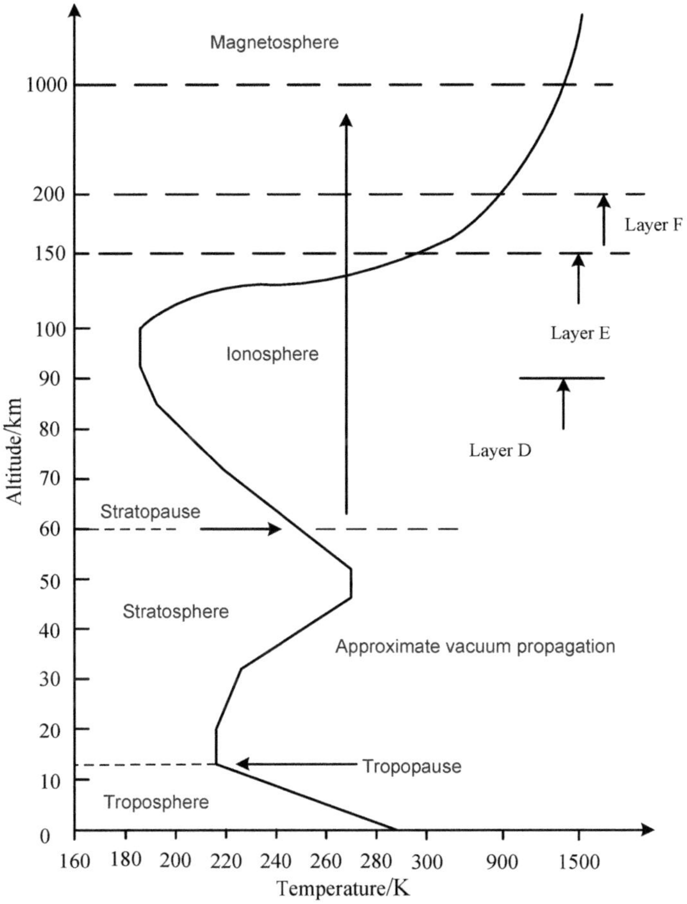

Fig. 1.5 Structure sketch of layering of the atmosphere

1.4 Function of Radio Wave Propagation in Designing of a Satellite System

The intricate and varying atmospheric environment makes the influence of the atmosphere and the ground on satellite signals a complex matter. The influence of radio wave propagation on the signals also varies substantially with different locations of ground users. For the 3 GHz frequency band and below, the propagation effects produced by the ionosphere on the radio waves, such as scintillation, time delay, and dispersion play an important role in the performance of a satellite system, but such effects vary significantly with latitudes. For example, scintillation always occurs in low-latitude areas near the equator (around the south of China) and mainly at night. A strong scintillation process can result in a scintillation fading of 20 dB signal and

Table 1.3 Main propagation effects of a satellite system

Environment factor	Influence mechanism for radio wave propagation	Relative frequency band
Troposphere	The scintillation effect due to inhomogeneities of the refractive index, including amplitude and phase of receiving signal, swift variation in angle of arrival, and reduction in effective gain of antennas	VHF/UHF/SHF/EHF
	Absorption of radio wave energy by gas molecules, such as oxygen and water vapor	UHF/SHF/EHF
	Absorption of radio wave by water vapor condensates (rain, snow, cloud, fog), attenuation, and depolarization generated by scattering	UHF/SHF/EHF
Ionosphere	Ionospheric refraction effect due to large- scale inhomogeneities of electron concentration, including group delay, dispersion, and Doppler frequency shift	VHF/UHF/SHF
	Scintillation due to small-scale inhomogeneities of electron concentration, including signal amplitude, phase, and random fluctuation in the angle of arrival	VHF/UHF/SHF
	Faraday rotation effect due to geomagnetic field, including polarization loss and reduction in cross-polarization resolution	VHF/UHF
	Ionospheric absorption due to electron collision	VHF
Ground (or sea) surface	Multipath decline and attenuation due to ground (or sea) surface reflection	VHF/UHF/SHF/EHF
	Multipath attenuation, shadowing attenuation, and diffraction attenuation due to surface features	VHF/UHF/SHF/EHF

above causing interruption of system links, but such scintillation is much less likely to occur around in northern China, and even if it takes place, the scintillation intensity is much smaller, producing a minimal influence on the system. When a satellite system operates at the 10 GHz frequency band and above, the meteorological environment becomes the primary factor affecting the radio wave propagation. However, the impact of rainfall is especially severe and could cause a signal attenuation of up to tens of decibels, or even a critical interruption of satellite link signals. Since China is spread over a vast territory crossing over different climatic regions (from wet and rainy southeastern coastal areas to dry and arid areas near the northwest frontier), there exist huge differences in features of radio wave propagation of the satellite systems. For users of mobile services provided by satellites, it is imperative to consider propagation effects, such as the multipath decline effect caused by ground (or sea) surface reflection, the shadowing effect of trees and buildings, and penetration loss of buildings, except for the atmosphere effect.

In view of the above-mentioned information, the influence of radio wave propagation, which has significant geographic differences, forms the essential factor to be considered for the scientific planning and design of a satellite system, directly influencing its reliability. Propagation design not only emerges as a crucial assurance of

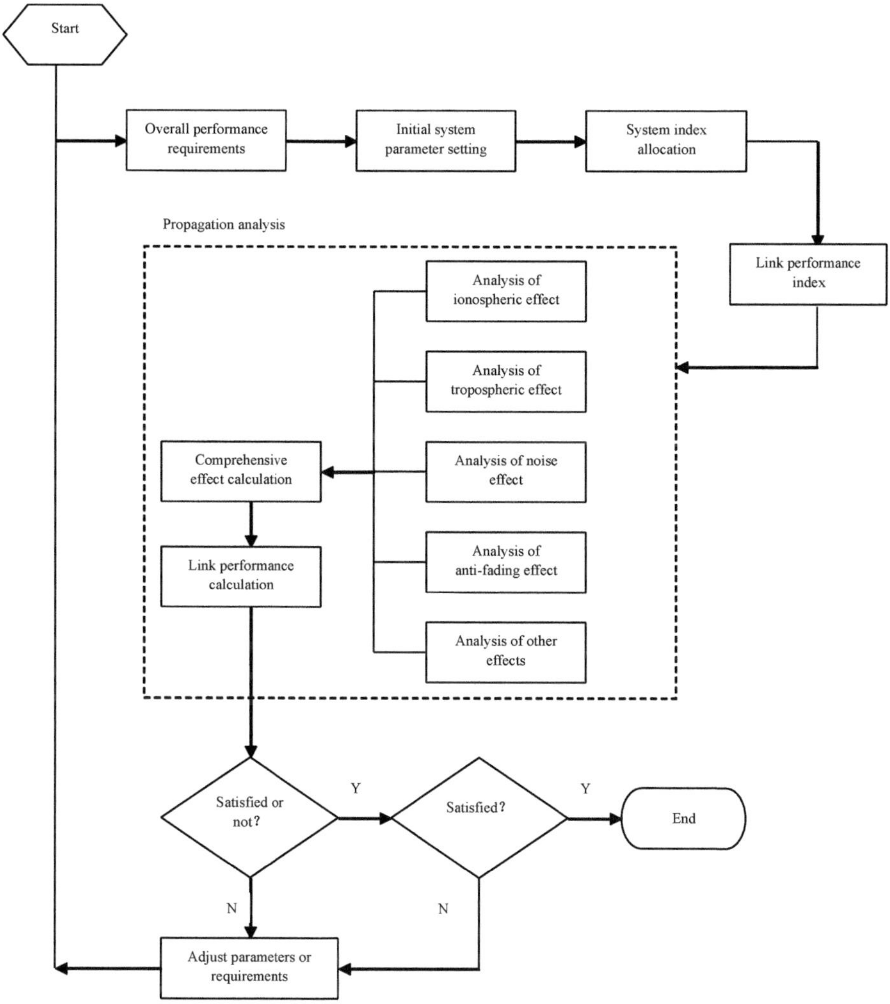

Fig. 1.6 Analysis process of common links

reliable operation under different application conditions of a satellite system as well as a key for its environmental suitability design, but also contributes greatly to the rational allocation of system resources of satellite and ground applications, enhances system performance, and offers technical support for the fabrication, development, and application of a satellite system [10–13].

The analysis process of common links in designing a satellite system is provided in Fig. 1.6.

References

1. Roddy D (2002) Satellite communication (Translation by Gengxin Z et al). Posts and Telecom Press, Beijing
2. Maral G, Bousquet M (2010) Satellite communications systems-systems, techniques and technology, 3rd edn. Wiley, US
3. Barclay L (ed) (2013) Propagation of radiowaves,3rd edn. The Institution of Engineering and Technology, London
4. Zhenwei Z (2001) Study on radio wave propagation characteristics and remote sensing of water condensates. Xidian University, Xi'an
5. Leke L (2004) Research on impact of tropospheric environment on earth-space propagation characteristics. Beijing. http://d.wanfangdata.com.cn/thesis/Y424844. China institute of electronics. http://d.wanfangdata.com.cn/thesis/y612399
6. Peixuan S, Jietai MAO, Jianguo L (2003) Atmospheric physics. Peking University Press, Beijing
7. Junxi G (1994) Dictionary of atmospheric science. China Meteorological Press, Beijing
8. Yixi X (2008) Principle and application of radio wave propagation. Posts and Telecommunications Press, Beijing
9. Jie H (1999) Correction of atmospheric refraction error of radio waves. National Defense Industry Press, Beijing
10. Zhenguo C, Hongwen Y, Wenbin G (2003) Satellite communication system and technology. Beijing University of Posts and Telecommunications Press, Beijing
11. Naitong Z, Zhongzhao Z, Yingtao L (2000) Satellite mobile communication system. Publishing House of Electronics Industry, Beijing
12. Liu Yuanjian Wu, Tao PS et al (2010) Introduction to modern communication technology. National Defense Industry Press, Beijing
13. Guoliang L, Kunbi R (1990) Satellite communication. Xidian University Press, Xi'an

Chapter 2
Clear-Air Effects

Tropospheric influence on the radio wave propagation of a satellite system can be primarily divided into clear-air effects and hydrometeor effects, which affect the key propagation factors determinative of the satellite's system performance at 30 GHz frequency band and above [1–3]. Clear-air effects mainly include refraction/scintillation and defocusing effects of altitude-varying atmosphere, the effect of runoff generation structure in the atmosphere on the radio waves, and absorption attenuation of the radio waves due to the oxygen and water vapor molecules present in the atmosphere [4, 5]. Hydrometeor effects are principal indicators of scattering/attenuation and depolarization effects generated at the time of interaction of radio waves with hydrometeors, such as rain, snow, cloud, and fog, which are the critical contributors affecting the performance of a satellite system operating at Ku frequency band and above. Troposphere propagation shares an inextricable link to meteorology and relevant climatic conditions. China possesses a widespread territory crossing into different climatic regions and hence, a great variety of effects of rainfall/cloud and gaseous attenuation on the radio signals of a satellite system in different regions can be observed. Therefore, according to the characteristics of radio waves in different environments, reliable forecasting of radio wave propagation in diverse regions becomes highly valuable in terms of applicability, design reliability, system resource allocation optimization, and satellite system management. The relationship between meteorological parameters of the radio waves of the troposphere and earth-space propagation effects is provided in Fig. 2.1 [6].

Z. Zhao et al., *Radio Wave Propagation of Satellite Systems*,
https://doi.org/10.1007/978-981-97-1394-3_2

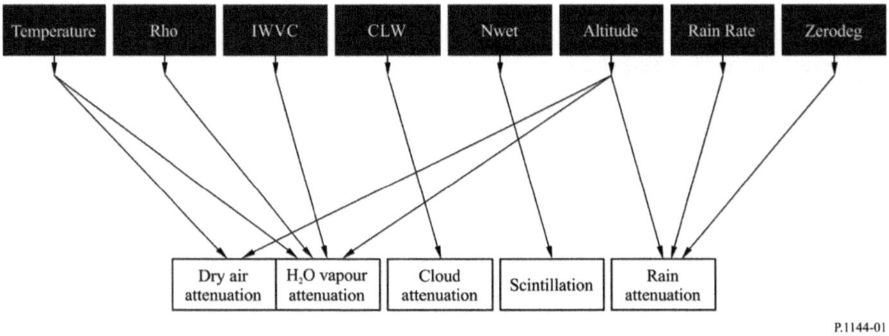

Fig. 2.1 The relationship between the geophysical maps (black boxes) and propagation effects (white boxes)

2.1 Atmospheric Refractive Effects

2.1.1 Atmospheric Refractive Index [7]

The ratio of the propagation speed of radio waves in a vacuum to that in optical media is called as media refractive index "*n*," which can be expressed using the following formula:

$$n = 1 + N \times 10^{-6} \tag{2.1}$$

In the formula: *N* denotes atmospheric refractivity, called as *N* unit, which can be expressed as a sum of dry and wet items of refraction rate:

$$N = N_{\text{dry}} + N_{\text{wet}} = \frac{77.6}{T}\left(P + 4810\frac{e}{T}\right) \quad \text{(N-units)} \tag{2.2}$$

In the formula: N_{dry} refers to a dry item of refractivity, $N_{\text{dry}} = 77.6 \times \frac{P}{T}$; N_{wet} refers to a wet item of refractivity, $N_{\text{wet}} = 3.732 \times 10^5 \times e/T^e$; *P* is atmospheric pressure (HPa); *e* represents vapor pressure (hPa); and *T* refers to absolute temperature (K). The relationship between vapor pressure "*e*" and relative humidity *H* (%):

$$e = \frac{H \cdot e_s}{100} \tag{2.3}$$

$$e_s = EF \cdot a \cdot \exp\left[\frac{\left(b - \frac{t}{d}\right) \cdot t}{t + c}\right] \tag{2.4}$$

Table 2.1 Value of parameters a, b, c, and d in different environments

Parameter	Water (–40 to +50 °C)	Ice (–80 to 0 °C)
a	6.1121	6.1115
b	18.678	23.036
c	257.14	279.82
d	234.5	333.7

In Formula (2.4): when calculating water, $EF_{\text{water}} = 1 + 10^{-4} \times [7.2 + P(0.0032 + 5.9 \times 10^{-7} \times t^2)]$ [1]; when calculating ice, $EF_{\text{ice}} = 1 + 10^{-4} \times [2.2 + P(0.00382 + 6.4 \times 10^{-7} \times t^2)]$.

In the formula: t is temperature (°C), e_s denotes saturated water vapor pressure (hPa) at temperature t. Parameters a, b, c, and d are listed in Table 2.1.

Vapor pressure e (hPa) can also be calculated using the following formula:

$$e = \frac{\rho T}{216.7} \tag{2.5}$$

In the formula: ρ signifies vapor density (g/m^3).

ITU-Radiocommunication Sector (ITU-R)-analyzed global statistical data for atmospheric refractivity at different altitudes for a period of 5 years collected from over 1,000 surface meteorological stations. The result of the analysis suggests that the atmospheric refractivity approximately varies linearly with the altitude from sea level up to 1 km. In the entire troposphere, the refractivity basically abides by the negative exponential distribution:

$$N_s = N_0 \exp\left(-h_s / h_0\right) \tag{2.6}$$

In the formula: N_s refers to atmospheric refractivity at altitude h_s; h_s denotes altitude (km); h_0 is nominal height (km); and N_0 refers to atmospheric refractivity at sea level. Under standard atmospheric conditions [8]: $N_0 = 315$, $h_0 = 7.35$ km.

2.1.2 Ray Bending and Effective Earth Radius [9]

Atmospheric refractive index gradient varies with the increase in altitude, resulting in the bending of radio waves when crossing through the troposphere. Since the refractive index decreases progressively with the increasing altitude by an exponential distribution, the vertical gradient of the refractive index must be taken into account when the refraction is considered. At this moment, the curvature of any point of radio rays can be expressed as

$$\frac{1}{\rho} = -\frac{\cos\varphi}{n}\frac{dn}{dh} \tag{2.7}$$

In the formula: ρ refers to curvature radius of ray path; n is refractive index of the atmosphere; $\frac{dn}{dh}$ represents the vertical gradient of refractive index; φ refers to elevation angle of ray at this point.

When the rays bend toward earth's surface, that is, when the refractive index reduces with altitude, the curvature of radio rays obtains a positive value. At radio wave frequencies, such bending effect or refraction appears independent of frequency. When the propagation path is close to the horizontal path, φ approximately equals to 0, and n is very close to 1, then Formula (2.7) can be further simplified as shown below:

$$\frac{1}{\rho} = -\frac{dn}{dh} \tag{2.8}$$

According to Formula (2.8), if the refraction gradient is a constant, then the propagation trace is an arc. At this moment, if the refractive index of the atmosphere is distributed evenly and horizontally, the propagation can be regarded as a rectilinear propagation by equalizing the earth to derive an effective earth's radius as $R_e = ka$. The relationship between effective earth radius and the real earth is

$$\frac{1}{ka} = \frac{1}{a} + \frac{dn}{dh} = \frac{1}{R_e} \tag{2.9}$$

In the formula: "a" refers to radius of real earth (km); k refers to effective earth radius factor.

Applying this method (effective earth radius), the path of propagation rays can be deemed as straight lines, regardless of their elevation angle. Noticeably, the refraction gradient can only be interpreted as a constant if and only if the propagation path is horizontal. Essentially, when the altitude is lower than 1 km, the profile of the refractive index can be approximated using a linear model. Under standard atmospheric conditions, $\frac{dN}{dh} = -39$ (N unit/km), effective earth's radius factor $k = 4/3$, effective earth's radius $R_e = 8{,}500$ km.

2.1.3 Apparent Elevation Angle of a Satellite

During practical implementation, the radio waves transmitted from the ground station bend toward earth's surface by reason of atmospheric refraction. The apparent elevation angle of the satellite is different from its real elevation angle in free-space propagation; the error of elevation angle $T(h, \theta)$ caused by the atmospheric refraction can be defined using the following formula:

$$\tau(h, \theta) = -\int_h^{\infty} \frac{n'(x)}{n(x)\tan\varphi} dx \tag{2.10}$$

In Formula (2.10), φ is determined by considering Snell's law in the polar coordinate system:

$$\cos\varphi = \frac{(r+h)n(h)\cos\theta}{(r+x)n(x)} \tag{2.11}$$

In the formula: h refers to the altitude of ground station (km); r denotes earth's radius (usually using 6,370 km); x represents the integration variable (km) in Formula (1.2); θ refers to elevation angle (°); and $n(x)$ and $n(h)$ refer to refractive indices at altitudes x and h, respectively.

ITU-R provides the approximation method for the determination of $\tau(h, \theta)$ under reference atmospheric conditions:

$$\tau(h,\theta) = \frac{1}{\left[1.314 + 0.6437\theta + 0.02869\theta^2 + h\left(0.2305 + 0.09428\theta + 0.01096\theta^2\right) + 0.008583h^2\right]} \tag{2.12}$$

The above formula is obtained in the case of $0 \le h \le 3$ km and $\theta_m \le \theta \le 10°$, but the case of $10° \le \theta \le 90°$ offers a remarkably approximate result. θ_m is the angle at which the radio waves touch the earth's surface at a tangent. θ_m can be calculated using the following formula:

$$\theta_m = -\arccos\left(\frac{r}{r+h} \cdot \frac{n(0)}{n(h)}\right) \tag{2.13}$$

An approximate value of θ_m can also be obtained using $\theta_m = -0.875\sqrt{h}$.

Assume the elevation angle of the satellite to be θ_0 and the minimum elevation angle of the ground station to be θ_m in the case of free-space propagation, and the angle correction corresponding to θ_m is τ (h, θ_m). The satellite can be positioned within the visual scope of the ground station only when the following formula is established:

$$\theta_m - \tau(h, \theta_m) \le \theta_0 \tag{2.14}$$

When the satellite is positioned within the visual scope of the ground station, the apparent elevation angle with consideration of atmospheric refraction can be derived using the following formula:

$$\theta - \tau(h, \theta) = \theta_0 \tag{2.15}$$

Here, the apparent elevation angle can also be calculated directly using the following formula:

$$\theta = \theta_0 + \tau_s(h, \theta_0) \tag{2.16}$$

Table 2.2 Angle deviation when radio waves pass through the atmosphere

Pitch angle/(°)	Average angle refraction error $\Delta\theta/(°)$			
	Polar continental climate zone	Temperate continental climate zone	Temperate marine climate zone	Tropical marine climate zone
1		–	–	
2	0.45	0.36	0.38	0.65
4	0.32	0.25	0.26	0.47
10	0.21	0.11	0.12	0.27
20	0.10	0.05	0.06	0.14
30		0.03	0.04	
Pitch angle/(°)	$\Delta\theta$ fluctuation per day			
1	0.1 (root-mean-square value)			
10	0.007 (root-mean-square value)			

In the formula: $\tau_s(h, \theta_0)$ is the function of θ_0.

$$\tau_s(h, \theta_0) = 1 \Big/ \left[1.728 + 0.5411\theta_0 + 0.03723\theta_0^2 + h(0.1815 + 0.06272\theta_0 + 0.0138\theta_0^2) + h^2(0.01727 + 0.008288\theta_0)\right] \tag{2.17}$$

Table 2.2 provides the average angle refraction error and random fluctuation value when the radio waves measured by radar, radio altimeter, and radio astronomy pass through the atmosphere.

2.1.4 *Focusing and Defocusing Effects*

In practice, antenna beams are widened or narrowed since the degree of refraction varies with the elevation angle of the antenna beams under the influence of the atmosphere, effectuating in varying gain of antennas as well as level of signal reception, but it counts for a negligible effect in case the elevation angle is larger than 3°. The following formula can be used to calculate the focusing and defocusing loss or gain of the atmosphere:

$$b = \pm \lg(B) \tag{2.18}$$

In which,

$$B = 1 - \left[0.5411 + 0.0744\theta_0 + h(0.06272 + 0.0276\theta_0) + 0.08288h^2\right] /$$

$$[1.728 + 0.5411\theta_0 + 0.0372\theta_0^2 + h(0.1815 + 0.06272\theta_0 + 0.0138\theta_0^2)$$
$$+h^2(0.01727 + 0.00828\theta_0)]^2$$

In the formula: θ_0 is the elevation angle between the receiving and launching terminal (°), $\theta_0 < 10°$; h is the altitude of the launching station or receiving station near the ground ($h < 3$ km); and b is the level fluctuation of the signal with respect to free-space propagation (dB). If the launching station is located on or close to the earth's surface, then the symbol in the formula should be "–," and if the satellite is the launching source, then the symbol should be "+."

2.1.5 Excess Radio Path Length

The refractive index of the troposphere is larger than 1 and varies with the altitude so that the radio wave propagation speed remains lower than the free-space speed of light. The length of the radio path between the ground station and the satellite exceeds the length of its geometric path since the propagation rays bend undergoing an atmospheric refraction effect. These differences in distance can be calculated using the following integrations:

$$\Delta L = \int_A^B (n - 1)ds \tag{2.19}$$

In the formula: s refers to the length of the propagation path; n refers to the refractive index; A and B refer to terminals of path.

When the temperature of earth's surface, atmospheric pressure, and relative temperature are known, the excess radio path length can be calculated using the following semi-rational formula:

$$\Delta L = \frac{\Delta L_V}{\sin\phi_0(1 + k\cot^2\phi_0)^{1/2}} + \delta(\phi_0, \Delta L_V) \tag{2.20}$$

In the formula: ϕ_0 refers to the elevation angle of the observation point; ΔL_V refers to the distance refraction error of the vertical path; k and δ (ϕ_0, ΔL_V) are correction items in exponential atmospheric model calculation.

Note that item δ (ϕ_0, ΔL_V) denotes the refraction bending effect of the rays, the value of which is usually negligible owing to its small value, except in the case of a small elevation angle. For example, the error is only 3.5 cm when the elevation angle is 10°, the error is only 0.1 mm when the elevation angle is 45°, but this item is not ignored when the elevation angle is smaller than 5°. Vertical distance error is determined using the following formula:

Table 2.3 Value of coefficient a_1 and b_1

Geographic position	a_1	b_1
Coastal region (islands or regions that are less than 10 km from coastline)	5.5×10^{-4}	2.91×10^{-2}
Equatorial inland region	6.5×10^{-4}	2.73×10^{-2}
Other regions	7.3×10^{-4}	2.35×10^{-2}

$$\Delta L_V = 0.00227P + f(T)H$$

In the formula: P refers to the atmospheric pressure of ground (hPa); $f(T)$ refers to function of ground temperature; $f(T)$ is dependent on the geographic position and can be determined using the following formula:

$$f(T) = a_1 \cdot 10^{b_1 T}$$

In the formula: T refers to temperature (°C); values of coefficient a_1 and b_1 are relative to geographic position, as shown in Fig. 2.3 (Table 2.3).

Assume that the change of atmospheric refractive index N with altitude h is in exponential distribution:

$$N(h) = N_s \exp\left(-h/h_0\right)$$

In the formula: N_s refers to average value of refractive index of earth's surface; h_0 is

$$h_0 = 10^6 \frac{\Delta L_V}{N_s}$$

Then, k can be obtained using the following formula: $k = 1 - \left[\frac{n_s r_s}{n(h_0) r(h_0)}\right]$.

In the formula: n_s and $n(h_0)$ refer to refractive index values at earth's surface and at altitude h_0, respectively; r and $r(h_0)$ refer to the distance of ground and altitude h_0 from earth's core, respectively.

2.2 Attenuation Due to Gaseous Absorption [10]

Stimulated by radio waves, gas molecules transit from one energy state to another, and gas molecules absorb the energy of radio waves causing attenuation of the intensity of radio signals [10]. Oxygen and water vapor in the troposphere act as the primary absorbers of radio waves, with the resulting attenuation caused by the gaseous absorption showing an increasing trend with the rise in frequency, accompanied by a large amount of resonant absorption frequency. At 35 GHz frequency band

and below, oxygen has a series of absorption lines near 60 Hz to form an oxygen absorption band and exhibits an isolated absorption peak at 118.74 GHz. The water vapor absorption line is found at 22.3, 183.3, and 323.8 GHz. Gaseous absorption produces pressure broadening effect, namely, the line width of gaseous absorption broadens with the increase in atmospheric pressure. When the standard atmospheric pressure of the ground is 1013 mbar, the width of the absorption line is between 1 and 3 MHz. At the height of 20 km, the atmospheric pressure, however, is only 13 mbar, and the width of the absorption line seems significantly reduced to tens of kHz. There are tens of oxygen absorption lines within the 50–70 GHz frequency band. The atmospheric pressure reduces with the rise in altitude, hence, the width of absorption lines is fairly greater than the interval of absorption lines when the line is near the ground, and the width of absorption lines overlaps to form a continuous oxygen absorption band. Nonetheless, the width of absorption lines is considerably smaller than the interval of absorption lines at high altitude, and at this moment, the continuous oxygen absorption band at low altitude transforms into independent absorption lines.

2.2.1 Line-by-Line Calculation of Gaseous Attenuation

1. **Line-by-line calculation of gaseous attenuation**

Gaseous attenuation can accumulate resonance absorption lines of oxygen and water vapor following a line-by-line summation, which can be used to precisely calculate the gaseous attenuation rate (characteristic attenuation), using atmospheric temperature, relative humidity, and pressure parameters taking into account below 10 GHz non-resonance Debye absorption spectrum of oxygen, above 100 GHz nitrogen attenuation caused by atmospheric pressure, and additional continuous water vapor absorption found during the test. The gaseous attenuation rate γ is calculated using the following formula:

$$\gamma = \gamma_o + \gamma_w = 0.1820 f N''(f) \qquad \text{dB/km} \tag{2.21}$$

In the formula: γ_o refers to attenuation rate of dry air (including oxygen attenuation, nitrogen attenuation caused by atmospheric pressure, and non-resonance Debye attenuation) (dB/km); γ_w refers to vapor attenuation rate (dB/km); f denotes frequency (GHz).

$N''(f)$ refers to imaginary part of the complex refractive index related to frequency, that is

$$N''(f) = \sum_i S_i F_i + N''_D(f) \tag{2.22}$$

In the formula: S_i refers to intensity of ith absorption line; F_i refers to shape factor of ith absorption line.

The summation of all the lines is calculated using the Formula (2.22). $N''_D(f)$ refers to the contribution of nitrogen absorption due to atmospheric pressure and continuous absorption spectrum of Debye absorption spectrum to imaginary part of the refractive index.

$$S_i = \begin{cases} a_1 \times 10^{-7} p(300/T)^3 \exp[a_2(1 - 300/T)] & \text{oxygen} \\ b_1 \times 10^{-1} e(300/T)^{3.5} \exp[b_2(1 - 300/T)] & \text{water vapor} \end{cases} \tag{2.23}$$

In the formula: *P* refers to dry-air pressure (hPa); *e* denotes vapor pressure (hPa); total atmospheric pressure is represented by $P_{\text{tot}} = P + e_o$. *T* refers to temperature (K). Vapor pressure *e* can be calculated using vapor density ρ:

$$e = \frac{\rho T}{216.7} \tag{2.24}$$

While calculating the gaseous attenuation rate, profile data of *P*, *e*, and *T* measured locally must be utilized effusively. The standard atmospheric data can be referred to and utilized in the absence of local measurement data.

Shape factor F_i of ith absorption line is

$$F_i = \frac{f}{f_i}\left[\frac{\Delta f - \delta(f_i - f)}{(f_i - f)^2 + \Delta f^2} + \frac{\Delta f - \delta(f_i + f)}{(f_i + f)^2 + \Delta f^2}\right] \tag{2.25}$$

In the formula: f_i refers to frequency of absorption line; Δf refers to width of absorption line and is indicated as

$$\Delta f = \begin{cases} a_3 \times 10^{-4}(p\theta^{(0.8 - a_4)} + 1.1e\theta) & \text{oxygen} \\ b_3 \times 10^{-4}(p\theta^{b_4} + b_5 e\theta^{b_6}) & \text{water vapor} \end{cases} \tag{2.26}$$

Due to Doppler broadening effect, line width Δf is required to be corrected as

$$\Delta f = \begin{cases} \sqrt{\Delta f^2 + 2.25 \times 10^{-6}} & \text{oxygen} \\ 0.535\Delta f + \sqrt{0.217\Delta f^2 + \frac{2.1316 \times 10^{-12} f_i^2}{\theta}} & \text{water vapor} \end{cases} \tag{2.27}$$

$$\delta = \begin{cases} (a_5 + a_6\theta) \times 10^{-4}(p + e)\theta^{0.8} & \text{oxygen} \\ 0 & \text{water vapor} \end{cases} \tag{2.28}$$

The spectral line of oxygen attenuation is provided in Table 2.4, and the spectral line value of water vapor attenuation is provided in Table 2.5.

Nitrogen absorption prompted by atmospheric pressure and continuous absorption of dry air formed by Debye absorption can be given as

Table 2.4 Spectral line of oxygen attenuation [10]

f_i	a_1	a_2	a_3	a_4	a_5	a_6
50.474214	0.975	9.651	6.690	0.0	2.566	6.850
50.987745	2.529	8.653	7.170	0.0	2.246	6.800
51.503360	6.193	7.709	7.640	0.0	1.947	6.729
52.021429	14.320	6.819	8.110	0.0	1.667	6.640
52.542418	31.240	5.983	8.580	0.0	1.388	6.526
53.066934	64.290	5.201	9.060	0.0	1.349	6.206
53.595775	124.600	4.474	9.550	0.0	2.227	5.085
54.130025	227.300	3.800	9.960	0.0	3.170	3.750
54.671180	389.700	3.182	10.370	0.0	3.558	2.654
55.221384	627.100	2.618	10.890	0.0	2.560	2.952
55.783815	945.300	2.109	11.340	0.0	−1.172	6.135
56.264774	543.400	0.014	17.030	0.0	3.525	−0.978
56.363399	1331.800	1.654	11.890	0.0	−2.378	6.547
56.968211	1746.600	1.255	12.230	0.0	−3.545	6.451
57.612486	2120.100	0.910	12.620	0.0	−5.416	6.056
58.323877	2363.700	0.621	12.950	0.0	−1.932	0.436
58.446588	1442.100	0.083	14.910	0.0	6.768	−1.273
59.164204	2379.900	0.387	13.530	0.0	−6.561	2.309
59.590983	2090.700	0.207	14.080	0.0	6.957	−0.776
60.306056	2103.400	0.207	14.150	0.0	−6.395	0.699
60.434778	2438.000	0.386	13.390	0.0	6.342	−2.825
61.150562	2479.500	0.621	12.920	0.0	1.014	−0.584
61.800158	2275.900	0.910	12.630	0.0	5.014	−6.619
62.411220	1915.400	1.255	12.170	0.0	3.029	−6.759
62.486253	1503.000	0.083	15.130	0.0	−4.499	0.844
62.997984	1490.200	1.654	11.740	0.0	1.856	−6.675
63.568526	1078.000	2.108	11.340	0.0	0.658	−6.139
64.127775	728.700	2.617	10.880	0.0	−3.036	−2.895
64.678910	461.300	3.181	10.380	0.0	−3.968	−2.590
65.224078	274.000	3.800	9.960	0.0	−3.528	−3.680
65.764779	153.000	4.473	9.550	0.0	−2.548	−5.002
66.302096	80.400	5.200	9.060	0.0	−1.660	−6.091
66.836834	39.800	5.982	8.580	0.0	−1.680	−6.393
67.369601	18.560	6.818	8.110	0.0	−1.956	−6.475
67.900868	8.172	7.708	7.640	0.0	−2.216	−6.545
68.431006	3.397	8.652	7.170	0.0	−2.492	−6.600
68.960312	1.334	9.650	6.690	0.0	−2.773	−6.650

(continued)

Table 2.4 (continued)

f_i	a_1	a_2	a_3	a_4	a_5	a_6
118.750334	940.300	0.010	16.640	0.0	−0.439	0.079
368.498246	67.400	0.048	16.400	0.0	0.0	0.0
424.763020	637.700	0.044	16.400	0.0	0.000	0.000
487.249273	237.400	0.049	16.000	0.0	0.000	0.000
715.392902	98.100	0.145	16.000	0.0	0.000	0.000
773.839490	572.300	0.141	16.200	0.0	0.000	0.000
834.145546	183.100	0.145	14.700	0.0	0.000	0.000

$$N_D''(f) = fp\theta^2\left[\frac{6.14\times10^{-5}}{d\left[1+\left(\frac{f}{d}\right)^2\right]}+\frac{1.4\times10^{-12}p\theta^{1.5}}{1+1.9\times10^{-5}f^{1.5}}\right] \tag{2.29}$$

In the formula: d refers to the width coefficient in the Debye absorption spectrum.

$$d = 5.6\times10^{-4}p\theta^{0.8} \tag{2.30}$$

Figure 2.2 exhibits the gaseous attenuation rate at sea level and dry-air attenuation under the standard atmospheric conditions (with 1013 hPa of sea-level pressure, 15 °C of temperature, and 7.5 g/m^3 of vapor density).

Figure 2.3 provides the gaseous attenuation rates near the 60 GHz oxygen absorption band at different altitudes. The figure illustrates that multiple oxygen absorption lines above sea level combine to form a continuous absorption band attributable to the pressure broadening effect, but at higher altitude, separate single absorption lines are lucidly visible due to the narrowing width of the oxygen absorption line.

2. **Integration calculation of gaseous attenuation of earth-space slant path**

The propagation path of a satellite system traces through the atmosphere, and the gaseous attenuation rates vary with different altitudes.

The total gaseous attenuation achieved via the slant path when radio waves pass through the atmosphere is obtained by path integration of gaseous attenuation calculated using the line-by-line integration method mentioned above. When altitude of the ground station is h and elevation angle of the satellite is θ, then gaseous attenuation of satellite-ground links A (h, θ) is calculated using the following formula:

$$A(h,\theta) = \int_h^{\infty}\frac{\gamma(H)}{\sin\varphi}dH \tag{2.31}$$

In the formula: calculation formula for φ refers to Formula (2.11).

Figure 2.4 provides standard atmosphere and dry-air attenuation in the plumb direction, with a frequency interval of 1 GHz and a vertical integration step of 1 km.

Table 2.5 Data of spectral line of water vapor attenuation [10]

f_i	b_1	b_2	b_3	b_4	b_5	b_6
22.235080	0.1079	2.144	26.38	0.76	5.087	1.00
67.803960	0.0012	8.735	28.58	0.69	4.930	0.82
119.995940	0.0008	8.356	29.48	0.70	4.780	0.79
183.310091	2.4200	0.668	30.50	0.64	5.300	0.85
321.225644	0.0483	6.181	23.03	0.67	4.690	0.54
325.152919	1.4990	1.540	27.83	0.68	4.850	0.74
336.222601	0.0011	9.829	26.93	0.69	4.740	0.61
380.197372	11.5200	1.048	28.73	0.54	5.380	0.89
390.134508	0.0046	7.350	21.52	0.63	4.810	0.55
437.346667	0.0650	5.050	18.45	0.60	4.230	0.48
439.150812	0.9218	3.596	21.00	0.63	4.290	0.52
443.018295	0.1976	5.050	18.60	0.60	4.230	0.50
448.001075	10.3200	1.405	26.32	0.66	4.840	0.67
470.888947	0.3297	3.599	21.52	0.66	4.570	0.65
474.689127	1.2620	2.381	23.55	0.65	4.650	0.64
488.491133	0.2520	2.853	26.02	0.69	5.040	0.72
503.568532	0.0390	6.733	16.12	0.61	3.980	0.43
504.482692	0.0130	6.733	16.12	0.61	4.010	0.45
547.676440	9.7010	0.114	26.00	0.70	4.500	1.00
552.020960	14.7700	0.114	26.00	0.70	4.500	1.00
556.936002	487.4000	0.159	32.10	0.69	4.110	1.00
620.700807	5.0120	2.200	24.38	0.71	4.680	0.68
645.866155	0.0713	8.580	18.00	0.60	4.000	0.50
658.005280	0.3022	7.820	32.10	0.69	4.140	1.00
752.033227	239.6000	0.396	30.60	0.68	4.090	0.84
841.053973	0.0140	8.180	15.90	0.33	5.760	0.45
859.962313	0.1472	7.989	30.60	0.68	4.090	0.84
899.306675	0.0605	7.917	29.85	0.68	4.530	0.90
902.616173	0.0426	8.432	28.65	0.70	5.100	0.95
899.306675	0.0605	7.917	29.85	0.68	4.530	0.90
902.616173	0.0426	8.432	28.65	0.70	5.100	0.95
906.207325	0.1876	5.111	24.08	0.70	4.700	0.53
916.171582	8.3400	1.442	26.70	0.70	4.780	0.78
923.118427	0.0869	10.220	29.00	0.70	5.000	0.80
970.315022	8.9720	1.920	25.50	0.64	4.940	0.67
987.926764	132.1000	0.258	29.85	0.68	4.550	0.90
1780.0000	22,300.00	0.952	176.20	0.50	30.500	5.00

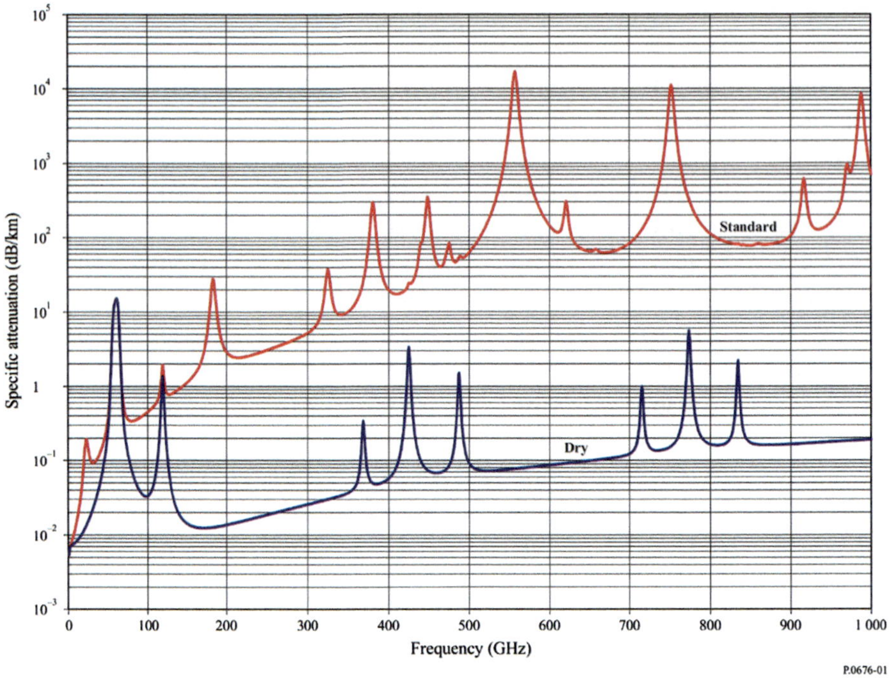

Fig. 2.2 Specific attenuation due to atmospheric gases [10]

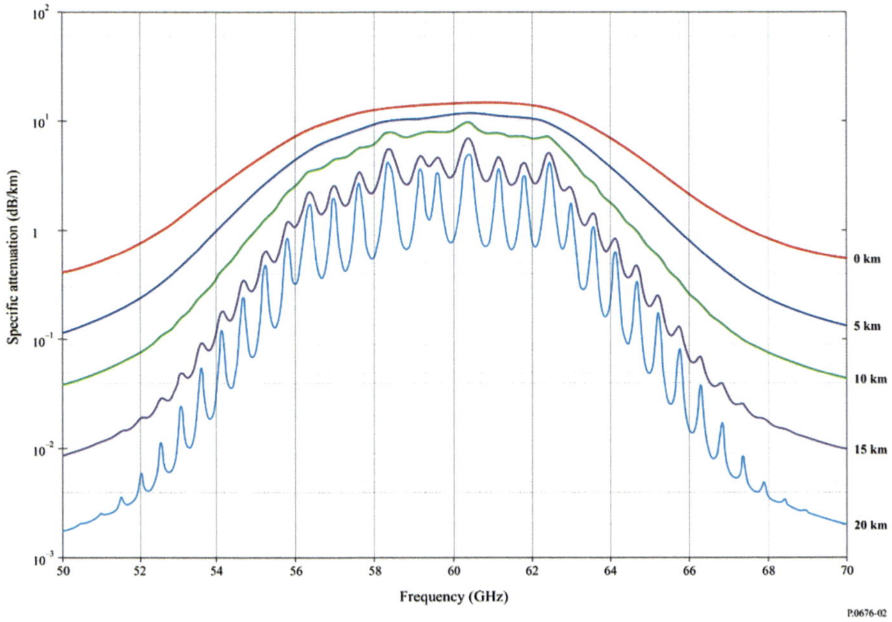

Fig. 2.3 Specific attenuation in the range 50–70 GHz at the different altitudes [10]

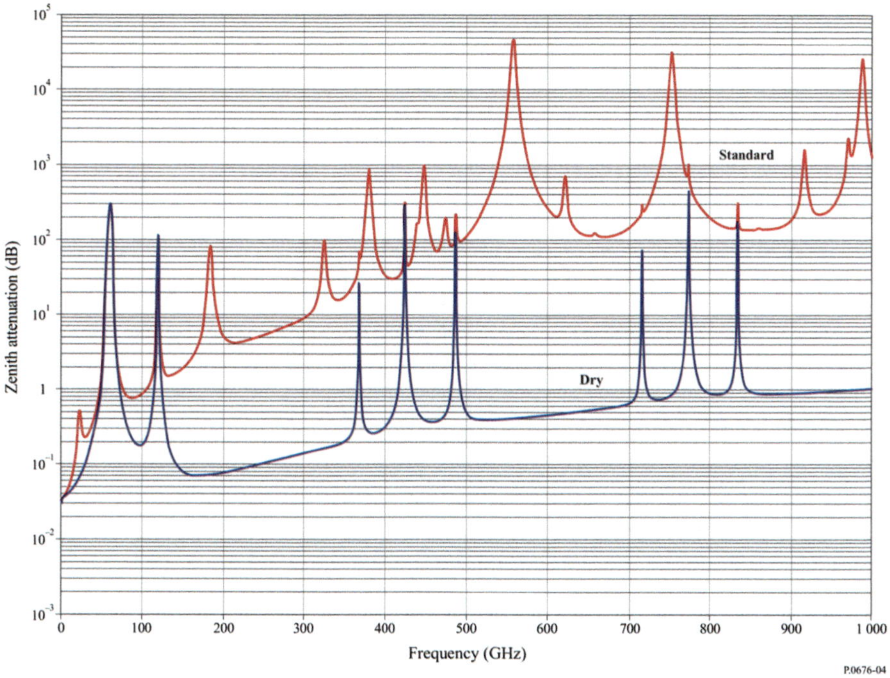

Fig. 2.4 Standard atmosphere and dry-air attenuation in the plumb direction [10]

A numerical algorithm, instead of integration formula, can be applied for measuring the gaseous attenuation via the slant path. Herein, the earth's atmosphere is layered by different thicknesses, bearing in mind the ray bending caused by each atmospheric refraction layer. As shown in Fig. 2.5, a_n refers to passing-through length of the electromagnetic waves in nth layer, δ_n refers to thickness of nth layer, n_n refers to refraction rate of nth layer, r_n denotes the distance from earth's core to the bottom of nth layer, α_n and β_n refer to angle of incidence and angle of emergence of nth layer. Gaseous attenuation via the slant path is calculated using the following formula:

$$A_{\text{gas}} = \sum_{n=1}^{k} a_n \gamma_n \tag{2.32}$$

In the formula: γ_n refers to the gaseous attenuation rate of nth layer calculated using the Formula (2.21); k represents the number of layers. In view of the relationship between vertical heterogeneity of atmosphere and altitude, the formula for thickness of nth layer δ_n (km) is as shown below:

$$\delta_n = 0.0001 \exp\left(\frac{n-1}{100}\right) \text{km} \tag{2.33}$$

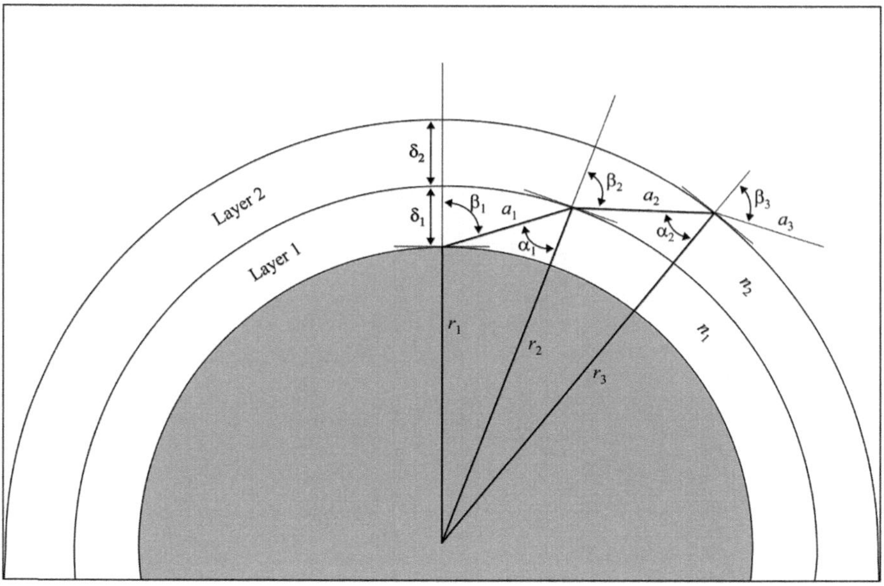

Fig. 2.5 Layering of gaseous attenuation calculation [10]

This indicates that the closer to the ground, the more precise is the layering; the thickness of the 1st layer is found to be 10 cm. For satellite-ground links, altitude up to 30 km can be considered for the calculation, while for oxygen absorption central frequency, altitude up to 100 km can be utilized. Calculation formula for a_n is as shown below:

$$a_n = -r_n \cos \beta_n + \frac{1}{2}\sqrt{4r_n^2 \cos^2 \beta_n + 8r_n\delta_n + 4\delta_n^2} \tag{2.34}$$

Calculation formula for α_n is as shown below:

$$\alpha_n = \pi - \arccos\left(\frac{-a_n^2 - 2r_n\delta_n - \delta_n^2}{2a_n r_n + 2a_n\delta_n}\right) \tag{2.35}$$

Calculation formula for β_{n+1} is as shown below:

$$\beta_{n+1} = \arcsin\left(\frac{n_n}{n_{n+1}} \sin \alpha_n\right) \tag{2.36}$$

$$\beta_1 = 90 - \theta$$

In the formula: θ refers to the elevation angle of the satellite from the ground station.

2.2.2 Simplified Method for Calculation of Gaseous Attenuation

1. Simplified approach of gaseous specific attenuation

Line-by-line calculation of gaseous specific attenuation seems incompatible with practical use owing to its complexity. Therefore, Zhang Minggao, an academician from the China Institute of Radiowave Propagation Research, presented a simplified approach for calculating the gaseous specific attenuation, which has been adopted by ITU-R. This simplified method is employed to calculate the gaseous specific attenuation at a frequency not more than 350 GHz and from sea level up to 10 km altitude. This method is consistent with the line-by-line calculation result, reporting a calculation error not more than 0.1 dB/km, and with a maximum error of 0.7 dB/km in the 60 GHz oxygen absorption band. When altitude exceeds 10 km, adopting line-by-line calculation becomes imperative to ensure calculation accuracy since the value of gaseous specific attenuation is small. The calculation process of simplified method of gaseous attenuation is as shown below:

Specific attenuation of dry air γ_o (dB/km) is calculated below:

When frequency $f \leq 54$ GHz, that is

$$\gamma_o = \left[\frac{7.2 r_t^{2.8}}{f^2 + 0.34 r_p^2 r_t^{1.6}} + \frac{0.62\xi_3}{(54 - f)^{1.16\xi_1} + 0.83\xi_2}\right] f^2 r_p^2 \times 10^{-3} \tag{2.37}$$

When frequency 54 GHz $< f \leq 60$ GHz, that is

$$\gamma_o = \exp\left[\frac{\ln\gamma_{54}}{24}(f-58)(f-60) - \frac{\ln\gamma_{58}}{8}(f-54)(f-60) + \frac{\ln\gamma_{60}}{12}(f-54)(f-58)\right] \tag{2.38}$$

When frequency 60 GHz $< f \leq 62$ GHz, that is

$$\gamma_o = \gamma_{60} + (\gamma_{62} - \gamma_{60})\frac{f-60}{2} \tag{2.39}$$

When frequency 62 GHz $< f \leq 66$ GHz, that is

$$\gamma_o = \exp\left[\frac{\ln\gamma_{62}}{8}(f-64)(f-66) - \frac{\ln\gamma_{64}}{4}(f-62)(f-66) + \frac{\ln\gamma_{66}}{8}(f-62)(f-64)\right] \tag{2.40}$$

When frequency 66 GHz $< f \leq 120$ GHz, that is

$$\gamma_o = \left\{3.02 \times 10^{-4} r_t^{3.5} + \frac{0.283 r_t^{3.8}}{(f - 118.75)^2 + 2.91 r_p^2 r_t^{1.6}} + \right.$$

$$\left.\frac{0.502\xi_6[1 - 0.0163\xi_7(f - 66)]}{(f - 66)^{1.4346\xi_4} + 1.15\xi_5}\right\} f^2 r_p^2 \times 10^{-3} \tag{2.41}$$

When frequency 120 GHz $< f \leq$ 350 GHz, that is

$$\gamma_o = \left[\frac{3.02 \times 10^{-4}}{1 + 1.9 \times 10^{-5} f^{1.5}} + \frac{0.283 r_t^{0.3}}{(f - 118.75)^2 + 2.91 r_p^2 r_t^{1.6}}\right] f^2 r_p^2 r_t^{3.5} \times 10^{-3} + \delta \tag{2.42}$$

In which,

$$\begin{aligned}
\xi_1 &= \varphi(r_p, r_t, 0.0717, -1.8132, 0.0156, -1.6515) \\
\xi_2 &= \varphi(r_p, r_t, 0.5146, -4.6368, -0.1921, -5.7416) \\
\xi_3 &= \varphi(r_p, r_t, 0.3414, -6.5851, 0.2130, -8.5854) \\
\xi_4 &= \varphi(r_p, r_t, 0.0112, 0.0092, -0.1033, -0.0009) \\
\xi_5 &= \varphi(r_p, r_t, 0.2705, -2.7192, -0.3016, -4.1033) \\
\xi_6 &= \varphi(r_p, r_t, 0.2445, -5.9191, 0.0422, -8.0719) \\
\xi_7 &= \varphi(r_p, r_t, -0.1833, 6.5589, -0.2402, 6.131) \\
\gamma_{54} &= 2.192\varphi(r_p, r_t, 1.8286, -1.9487, 0.4051, -2.8509) \\
\gamma_{58} &= 12.59\varphi(r_p, r_t, 1.0045, 3.5610, 0.1588, 1.2834) \\
\gamma_{60} &= 15\varphi(r_p, r_t, 0.9003, 4.1335, 0.0427, 1.6088) \\
\gamma_{62} &= 14.28\varphi(r_p, r_t, 0.9886, 3.4176, 0.1827, 1.3429) \\
\gamma_{64} &= 6.819\varphi(r_p, r_t, 1.4320, 0.6258, 0.3177, -0.5914) \\
\gamma_{66} &= 1.908\varphi(r_p, r_t, 2.0717, -4.1404, 0.4910, -4.8718) \\
\delta &= -0.00306\varphi(r_p, r_t, 3.211, -14.94, 1.583, -16.37) \\
\varphi(r_p, r_t, a, b, c, d) &= r_p^a r_t^b \exp[c(1 - r_p) + d(1 - r_t)]
\end{aligned}$$

In the formula: f refers to frequency (GHz); $r_p = p/1013$; $r_t = 288/(273 + t)$; p refers to pressure (hPa); t refers to temperature (°C).

Specific attenuation of water vapor γ_w (dB/km) is calculated below:

$$\gamma_w = \left\{\frac{3.98\eta_1 \exp[2.23(1 - r_t)]}{(f - 22.235)^2 + 9.42\eta_1^2} g(f, 22) + \frac{11.96\eta_1 \exp[0.7(1 - r_t)]}{(f - 183.31)^2 + 11.14\eta_1^2} + \right.$$
$$\frac{0.081\eta_1 \exp[6.44(1 - r_t)]}{(f - 321.226)^2 + 6.29\eta_1^2} + \frac{3.66\eta_1 \exp[1.6(1 - r_t)]}{(f - 325.153)^2 + 9.22\eta_1^2} +$$
$$\frac{25.37\eta_1 \exp[1.09(1 - r_t)]}{(f - 380)^2} + \frac{17.4\eta_1 \exp[1.46(1 - r_t)]}{(f - 448)^2} +$$

$$\frac{844.6\eta_1 \exp[0.17(1-r_t)]}{(f-557)^2} g(f,557) +$$
$$\frac{290\eta_1 \exp[0.41(1-r_t)]}{(f-752)^2} g(f,752) +$$
$$\left.\frac{8.3328\times10^4\eta_2 \exp[0.99(1-r_t)]}{(f-1780)^2} g(f,1780)\right\} f^2 r_t^{2.5} \rho \times 10^{-4} \quad (2.43)$$

In the formula:

$$\eta_1 = 0.955 r_p r_t^{0.68} + 0.006\rho$$
$$\eta_2 = 0.735 r_p r_t^{0.5} + 0.0353 r_t^4 \rho$$
$$g(f,f_i) = 1 + \left(\frac{f-f_i}{f+f_i}\right)^2$$

2. **Simplified calculation of gaseous attenuation of earth-space slant path**

Atmospheric absorption attenuation can be calculated conveniently by using vapor equivalent altitude and oxygen equivalent altitude. This simplified calculation appears quite suitable for the case when the elevation angle is between 5° and 90°, and the altitude is not more than 10 km.

The simplified calculation formula for gaseous attenuation taking place in an earth-space slant path based on meteorological data of earth's surface is as shown below:

$$A = \frac{A_0 + A_w}{\sin\theta} = \frac{h_0\gamma_0 + h_w\gamma_w}{\sin\theta} \quad (2.44)$$

In the formula: θ is elevation angle (°); γ_o refers to specific attenuation due to oxygen (dB/km); γ_w refers to specific attenuation due to water vapor (dB/km); h_o refers to oxygen equivalent altitude (km); h_w refers to vapor equivalent altitude (km); $A_0 = h_0\gamma_0$ refers to oxygen attenuation in the zenith direction (dB); $A_w = h_w\gamma_w$ refers to water vapor attenuation in the zenith direction (dB).

Calculation formula for h_o is as shown below:

$$h_o = \frac{6.1}{1 + 0.17 r_p^{-1.1}}(1 + t_1 + t_2 + t_3) \quad (2.45)$$

In which,

$$t_1 = \frac{4.64}{1 + 0.066 r_p^{-2.3}} \exp\left[-\left(\frac{f-59.7}{2.87 + 12.4\exp(-7.9 r_p)}\right)^2\right]$$
$$t_2 = \frac{0.14\exp(2.12 r_p)}{(f-118.75)^2 + 0.031\exp(2.2 r_p)}$$

$$t_3 = \frac{0.0114}{1 + 0.14 r_p^{-2.6}} f \frac{-0.0247 + 0.0001 f + 1.61 \times 10^{-6} f^2}{1 - 0.0169 f + 4.1 \times 10^{-5} f^2 + 3.2 \times 10^{-7} f^3}$$

Formula (2.45) should satisfy with the constraint condition: when $f < 70$ GHz, $h_o \le 10.7 r_p^{0.3}$.

Calculation formula for h_w is as shown below:

$$h_w = 1.66 \left(1 + \frac{1.39\sigma_w}{(f - 22.235)^2 + 2.56\sigma_w} + \frac{3.37\sigma_w}{(f - 183.31)^2 + 4.69\sigma_w} + \frac{1.58\sigma_w}{(f - 325.1)^2 + 2.89\sigma_w} \right) \tag{2.46}$$

$$\sigma_w = \frac{1.013}{1 + \exp[-8.6(r_p - 0.57)]}$$

Calculation formula for the accumulation distribution of gaseous attenuation of the earth-space slant path based on the accumulation distribution of total columnar water vapor found in the zenith direction is as shown below:

$$A_{gas}(p) = \frac{A_0 + A_w(p)}{\sin\theta} \tag{2.47}$$

In the formula: $A_{gas}(p)$ refers to gaseous attenuation with $p\%$ time probability exceeded; $A_w(p)$ refers to vapor attenuation in the plumb direction with $p\%$ time probability exceeded; the calculation formula is as shown below:

$$A_w(p) = \frac{0.0173 V_t(p) \gamma_w(f, P_{ref}, \rho_{ref}, t_{ref})}{\gamma_w(f_{ref}, P_{ref}, \rho_{ref}, t_{ref})} \tag{2.48}$$

In the formula: f refers to frequency (GHz); $f_{ref} = 20.6$ (GHz); $P_{ref} = 780$ (hPa); $V_t(p)$ refers to integration vapor content with $p\%$ time probability exceeded; $\gamma_w(f, P_{ref}, \rho_{ref}, t_{ref})$ can be calculated using Formula (2.43); calculation formula for ρ_{ref} (g/m^3) and t_{ref} (°C) is shown as follows:

$$\rho_{ref} = \frac{V_t(p)}{4}$$

$$t_{ref} = 14 \ln\left(\frac{0.22 V_t(p)}{4} \right) + 3$$

2.3 Tropospheric Scintillation and Multipath Fading

Refractive index of troposphere decreases regularly with the altitude and causes refraction of radio waves inducing ray bending, however, small-scale inconsistencies of the refractive index trigger rapid fluctuation of amplitude and phase in electromagnetic waves, namely, the tropospheric scintillation effect. The measured result of one tropospheric scintillation event is provided in Fig. 2.6. Tropospheric scintillation effect is closely linked to the frequency and elevation angle of the system. Scintillation affects the satellite system marginally when the elevation angle is larger than 5° and frequency is smaller than 10 GHz; while it can generate severe deterioration of system performance when the elevation angle is smaller than 5° and frequency is higher than 10 GHz.

According to existing research, the amplitude of tropospheric scintillation is dependent on the change in amplitude and structure of atmospheric refractive index and is directly proportional to the frequency of the system, length of radio waves passing through the troposphere, and width of antenna beams. Based on substantial test data, scintillation can be known to exhibit significant seasonal characteristics since the meteorological environment varies considerably in different seasons. An intense scintillation event is always observed in the summer season. Scintillation may also undergo significant daily change, which is still smaller than the seasonal change. Moreover, the monthly average RMS value of scintillation fluctuation is highly related to the wet item of the refractive index. The index is determined by the vapor content in the atmosphere.

Scintillation fading of satellite-ground links with low elevation angle tends to be acute, and its fading features and fading depth distribution are analogous to multipath fading of ground links. At the same time, the fading distribution of satellite-ground

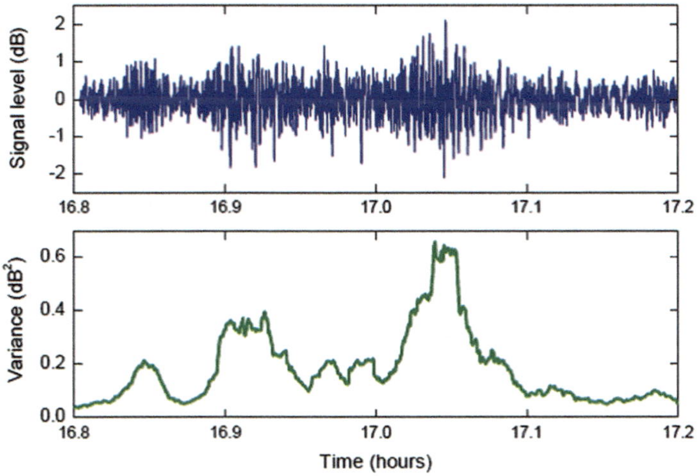

Fig. 2.6 Measured result of tropospheric scintillation

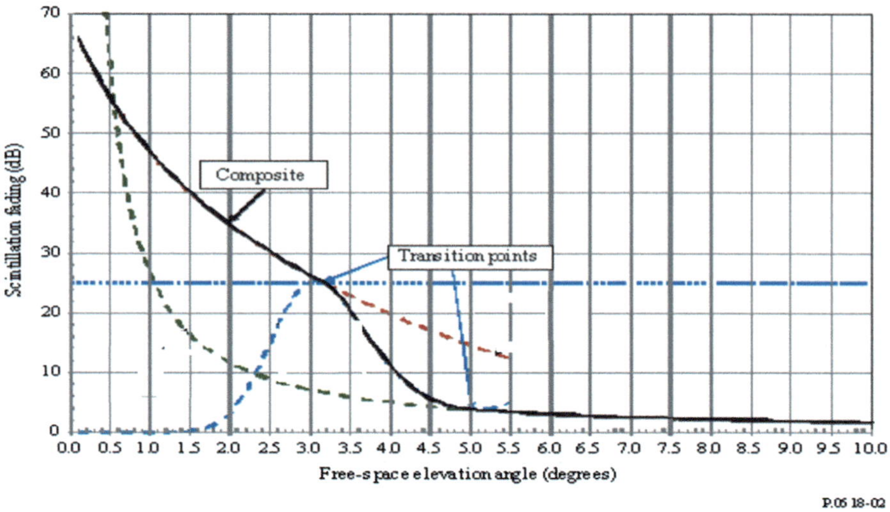

Fig. 2.7 Illustrative example of the scintillation prediction method of ITU-R [11]

links of low elevation angle also appears closely related to the statistical feature of refractive index gradient. According to the overall distribution of fading, it is considered that scintillation fading with a large time probability gradually transits to multipath fading distribution with a small time probability. In fact, clear-air fading distribution due to the atmospheric refraction effect is a combined distribution of bean expanding effect, scintillation effect, and multipath effect. ITU-R predicts the scintillation fading of the troposphere under three conditions, given as follows [11]:

① Elevation angle larger than 5°.
② Fading depth larger than 25 dB.
③ Transition zones.

Figure 2.7 demonstrates the forecasting result examples.

2.3.1 *Prediction of Scintillation Fading at Elevation Angles Greater than 5°*

This approach is based on monthly average temperature, long-term average temperature t (°C), and relative humidity H. Scintillation fading depth distribution exhibits seasonal changes only because the average value of t and of H change with seasons. Scintillation fading depth with p time probability exceeded $A_s(p)$ is calculated using the following formula:

$$A_S(p) = a(p) \cdot \sigma \tag{2.49}$$

In the formula: $a(p)$ $(0.01\% < p \leq 50\%)$ is time probability factor:

$$a(p) = -0.061(\log_{10} p)^3 + 0.072(\log_{10} p)^2 - 1.71 \log_{10} p + 3.0 \quad (2.50)$$

σ refers to standard deviation of signal, which can be defined as

$$\sigma = \sigma_{\text{ref}} f^{7/12} \frac{g(x)}{(\sin\theta)^{1.2}} \quad (2.51)$$

In which, f refers to frequency (GHz), σ_{ref} refers to standard deviation of signal amplitude, which can be defined as

$$\sigma_{\text{ref}} = 3.6 \times 10^{-3} + 10^{-4} \times N_{\text{wet}} \quad (2.52)$$

In the formula: N_{wet} refers to wet item of refractivity.

In Formula (2.51): $g(x)$ refers to the average factor of antennas, which can be defined as

$$g(x) = \sqrt{3.86(x^2 + 1)^{11/12} \cdot \sin\left[\frac{11}{6} \arctan \frac{1}{x}\right] - 7.08x^{5/6}} \quad (2.53)$$

$$x = 1.22 D_{\text{eff}}^2 (f/L) \quad (2.54)$$

In the formula: D_{eff} denotes the effective diameter of antennas, which can be defined as

$$D_{eff} = \sqrt{\eta} D \quad \text{m} \quad (2.55)$$

In the formula: n refers to antenna efficiency. If antenna efficiency is not known, $\eta = 0.5$ can be used conventionally; D refers to the physical diameter of antennas of the ground station (m); L refers to effective path length (m):

$$L = \frac{2h_L}{\sqrt{\sin^2\theta + 2.35 \times 10^{-4}} + \sin\theta} \quad \text{m} \quad (2.56)$$

In the formula: h_L refers to height of turbulent layer, which usually is $h_L = 1000$ m; θ refers to elevation angle of path (°).

Note that when value of root in Formula (2.54) is a negative number, the scintillation fading is considered to be 0.

2.3.2 Prediction of Scintillation Fading for Fade Depth Greater than 25 dB

Fading depth of more than 25 dB is deemed to be an inclusive propagation effect of beam expanding, scintillation, and multipath. Apparent elevation angle is θ (mrad), and frequency is f (GHz). Yearly average accumulation distribution and worst-month accumulation distribution of fading with $p\%$ time probability exceeded can be calculated using Formulas (2.57) and (2.58), respectively.

$$A_y(p) = 10\log_{10} K_w - \nu + 9\log_{10} f - 59.5\log_{10}(1+\theta) - 10\log_{10} p \tag{2.57}$$

$$A_m(p) = 10\log_{10} K_w + 9\log_{10} f - 55\log_{10}(1+\theta) - 10\log_{10} p \tag{2.58}$$

In the formula:

$$\nu = -1.8 - 5.6\log_{10}\left(1.1 \pm |\cos 2\psi|^{0.7}\right) \tag{2.59}$$

$$K_w = p_L^{1.5} 10^{0.1(C_0 + C_{\mathrm{Lat}})} \tag{2.60}$$

In Formula (2.60): sign "+" is taken when absolute value of latitude $|\psi| \leq 45°$, and sign "−" is taken when $|\psi| > 45°$. In Formula (2.61): p_L refers to worst-month probability of refractivity gradient less than −100 N (unit/km) at 100 m near ground; value of C_0 is listed in Table 2.6; latitude coefficient C_{Lat} is calculated using the following formula:

$$C_{\mathrm{Lat}} = \begin{cases} 0 & \psi \leq 53° \text{ N or S} \\ -53 + \psi & 53° \text{ N or S} \leq \psi \leq 60° \text{ N or S} \\ 7 & 60° \text{ N or S } \leq \psi \end{cases} \tag{2.61}$$

The frequency and elevation angle suitable for the above-mentioned prediction methods is 1–45 GHz and 0.5°–5°, respectively.

Table 2.6 Value of C_0 under different propagation path types [11]

Path type	C_0
Propagation path is completely in line with the land, and antenna height of ground station is below 700 m above sea level	76
The antenna height of ground station is 700 m above sea level	70
Propagation path is above or partially above water surface (including near-water coastal area), "r" refers to the percentage of the path over water or near-water coastal areas in the overall propagation path, but intersection of rivers and small lakes is not included in the water surface	76 + 6r

2.3.3 Prediction of Scintillation Fading in Transition Regions

Interpolating by applying the cubic exponential model to obtain a prediction formula for scintillation fading of transition region $A(p)$, with scintillation fading depth less than 25 dB and free-space elevation angle less than 5°:

$$A(p) = A_1 \exp\left[\alpha(p)(\theta - \theta_1) + \beta(p)(\theta - \theta_1)^2 + \gamma(p)(\theta - \theta_1)^2(\theta - \theta_2)\right] \quad (2.62)$$

where

$$\alpha(p) = \frac{A_1'}{A_1} \quad (2.63)$$

$$\beta(p) = \frac{\ln\left(\frac{A_2}{A_1}\right) - \alpha\delta}{\delta^2} \quad (2.64)$$

$$\gamma(p) = \frac{A_2' - A_2(\alpha + 2\beta\delta)}{A_2\delta^2} \quad (2.65)$$

$$\delta = \theta_2 - \theta_1 \quad (2.66)$$

In the formula: θ_1 refers to frequency f (GHz), time probability refers to P (%); corresponding apparent elevation angle when fading depth $A_1 = 25$ (dB) is calculated using the following formula (v and K_w refer to Formula (2.59) and Formula (2.60), respectively):

$$\theta_1 = \begin{cases} \left(\dfrac{K_w f^{0.9}}{p10^{\frac{A_1}{10}}}\right)^{\frac{1}{5.5}} - 1 & \text{worst month} \\ \left(\dfrac{K_w 10^{-\frac{v}{10}} f^{0.9}}{p10^{\frac{A_1}{10}}}\right)^{\frac{1}{5.95}} - 1 & \text{average year} \end{cases} \quad (2.67)$$

In the formula: θ_2 refers to the apparent elevation angle (mrad) considering the refraction effect produced when free-space elevation angle is 5°; A_2 refers to scintillation fading when free-space elevation angle is 5°.

Based on Formula (2.49), A_1' (dB/mrad) is calculated using the following formula:

$$A_1' = \begin{cases} -\frac{55}{1+\theta_1} \log_{10} e & \text{worst month} \\ -\frac{59.5}{1+\theta_1} \log_{10} e & \text{average year} \end{cases} \quad \text{dB/mrad} \quad (2.68)$$

Calculation formula for A_2' (dB/mrad) is as shown below:

$$A_2' = A_2 \times \left[\frac{g'(x)}{g(x)}\frac{dx}{d\theta} - \frac{1.2}{\tan(\theta)}\right] \times \frac{1}{1000} \quad \text{dB/mrad} \quad (2.69)$$

$$\frac{g'(x)}{g(x)} = \frac{1770(x^2+1) + 2123x^{\frac{1}{6}}(x^2+1)^{\frac{11}{12}}[\cos\zeta - x\sin\zeta]}{12x^{\frac{1}{6}}(x^2+1)\left[354x^{\frac{5}{6}} - 193(x^2+1)^{\frac{11}{12}}\sin\zeta\right]} \tag{2.70}$$

$$\frac{dx}{d\theta} = \frac{1.22D_{\text{eff}}^2 f}{2h_L}\left[\frac{\sin\theta}{\sqrt{\sin^2\theta + 2.35\times 10^{-4}}} + 1\right]\cos\theta \tag{2.71}$$

$$\zeta = \frac{11}{6}\tan^{-1}\frac{1}{x} \tag{2.72}$$

This prediction method is applicable to the apparent elevation angle of the transition region (for example $\theta_l \le \theta \le \theta_2$) and satisfies the condition of $0 \le p\% \le 50\%$.

References

1. Hao X et al (2000) Radio wave propagation. Publishing House of Electronics Industry, Beijing
2. Zhenwei Z. Study on radio wave propagation characteristics and remote sensing of hydrometeor. Xidian University, 200L. L-114
3. Leke L (2004) Study on the influence of tropospheric environment on the characteristics of air-ground propagation. China Institute of Electronics Science
4. Shaohua Y (1989) Characteristics of tropospheric scintillation at 12 GHz. Wuhan University
5. Ishimaru (1986) Wave propagation and scattering in random media. In: Runheng H et al. Science Press, Beijing
6. ITU-R Recommendation P.1144-12 (2023) Guide to the application of the propagation methods of radiocommunication study group 3 [s]. Geneva.1TU-R
7. ITU-R Recommendation P.453-10 (2012) The radio refractive index: its formula and refractivity data [s]. Geneva.1TU-R
8. ITU-R Recommendation P.835-5 (2012) Reference standard atmospheres [s]. Geneva. ITU-R
9. ITU-R Recommendation I P.834-8 (2016) Effects of tropospheric refraction on radiowave propagation [s].Geneva.ITU-R
10. ITU-R Recommendation P.676-11 (2016) Attenuation by atmospheric gases [s]. Geneva.1TU-R
11. ITU-R Recommendation P.618-12 (2015) Propagation data and forecasting methods required for the design of Earth-space telecommunication systems [s]. Geneva.1TU-R

Chapter 3
Hydrometeor Effects

Hydrometeors (cloud, rain, and fog) being the most predominant atmospheric phenomena, can produce additional attenuation and depolarization effect on radio wave, considerably effecting the performance of a satellite communication system operating at 10 GHz and above. Rain attenuation that fails to greatly affect the communication of C-band satellite may interrupt the communication of Ku-band satellite. Cloud attenuation that rarely affects the Ku-band may influence the Ka-band seriously. Rain may produce even more acute attenuation for Ka-band than for Ku-band; similarly, rainstorms can induce signal attenuation up to tens of dB, causing signal interruption. Scattering of rain drops and ice layer may cause similar frequency interference between the ground and the earth-space links, triggering a decrease in cross-polarization discrimination (XPD) or resolution of the cross-polarization resolution for multiplexing frequency system and additional delay of signal. Radio wave propagation of hydrometeors forms a complex random process, and its propagation characteristics are dependent on frequency, meteorological conditions, geographic position, and climatic regions. Reliable predicting of hydrometeor effects is the essential foundation for the link design of a satellite system as well as its functioning competency under various bad propagation conditions.

3.1 Physical Characteristics of Hydrometeors

3.1.1 Physical Characteristics of Clouds and Fog

Clouds and fog are a colloidal system essentially composed of suspended water droplets or ice particles, namely aerosols, slowly undergoing sedimentation in near-ground air. In terms of the size, the diameter of clouds and fog particles varies from tens to dozens of microns. The fog appears in the bottom layer and is considered to be the cloud touching the ground. According to the location and mechanism of fog

Z. Zhao et al., *Radio Wave Propagation of Satellite Systems*,
https://doi.org/10.1007/978-981-97-1394-3_3

formation, fog can be divided into two kinds, such as advection fog and radiation fog. In general, advection fog is the kind formed when warm air moves to the cold underlying surface. Sea fog is one of such fog types. Radiation fog is primarily caused by the cooling of surface radiation, and one of its types is inland fog. Water content in fog varies with the intensities of fog and is directly relative to temperature. For moderate-intensity fog, water content is usually about 0.1–0.2 g/m^3 at −20 to −15 °C, 0.2–0.5 g/m^3 at −15 to 0 °C, and up to 0.5–1 g/m^3 at 0–10 °C [1]. Clouds and fog share identical characteristics. According to the international classification method, clouds are divided into high clouds, medium clouds, low clouds, and vertical development clouds in accordance with the height, and these 4 families of clouds mainly consist of 10 types, as shown in Table 3.1. Water content in clouds is usually higher than that in fog, and the highest water content is found in cumulus clouds with an average of 2 g/m^3, or up to 25–30 g/m^3. The higher the water content, the greater the attenuation of radio signals, and the attenuation of clouds and fog increases with the rise in frequency.

As the drop size distribution of clouds and fog particles varies over a wide range, people often tend to adopt different distributions to best manifest the spectral characteristics of the drops in accordance with the diverse measured drop size distribution, in which, the simplified gamma size distribution of cloud-and-fog drop (Khragi-an-Mazin distribution calculation method) is widely used [2]:

$$n(r) = a_f r^2 \exp\left(-b_f r\right) \tag{3.1}$$

In the formula: r refers to radius of cloud-and-fog drop; $n(r)$ refers to the number of drops in an interval of unit volume and unit radius.

Unit of r corresponds to unit of $n(r)$, for example, unit of $n(r)$ is m^{-4} when unit of r is m; unit of $n(r)$ is m^{-3} μm^{-1} when unit of r is μm; the relationship between a_f, b_f, visibility V (km), and water content W (g/m^3) is shown as follows [2]:

Table 3.1 International classification of clouds [1]

Cloud family	Height/km			Cloud genera
	Polar region	Temperate zone	Tropical zone	
High clouds	3–4	5–13	6–18	Cirrus clouds (Ci), Cumulus clouds (Cc), Cirrostratus (Cs)
Medium clouds	2–4	2–7	2–18	Altocumulus clouds (Ac), Altostratus clouds (As)
Low clouds		Ground-2		Stratocumulus clouds (Sc), Nimbostratus clouds (Ns), Stratus clouds (St)
Vertical development clouds		Ground-2		Cumulus clouds (Cu), Cumulonimbus clouds (Cb)

$$a_f = \frac{9.781}{V^6 W^5} \times 10^{15} \tag{3.2}$$

$$b_f = \frac{1.304}{VW} \times 10^4 \tag{3.3}$$

3.1.2 Physical Characteristics of Rainfall

1. Physical characteristics of rainfall:

It is crucial to first analyze the scattering characteristics of rainfall toward radio waves before delving into the influence produced by rainfall on the radio wave propagation of a satellite system; the scattering characteristics depend on physical characteristics, such as size, shape, orientation, and falling speed of a raindrop. The shape of a raindrop is relative to its size. The diameter of a raindrop is generally between 0.1 and 8 mm, and if it exceeds that measure, the raindrop becomes unstable and can easily rupture. According to photographic studies, raindrops with a radius of less than 1 mm are found to be spherical. For a larger-sized raindrop, its shape is ellipsoidal at the top and has a groove at the bottom, as shown in Fig. 3.1, with its rotation axis being approximately vertical. For ease of calculation and comparison, normally the concept of the equi-volume sphere is utilized. Existing researches suggest that for the calculation of raindrop scattering characteristics, the ratio of the short axis to long axis instead of the precise shape of a raindrop emerges as the most indispensable parameter of raindrop characterization; hence, the raindrop is usually deemed as an oblate ellipsoid possessing different ratios of long to short axis. For an oblate ellipsoid-shape raindrop, its size is characterized by the ratio a_r/b_r of short axis a_r to long axis b_r and effective earth's radius a_0, in which the relationship between a_r/b_r and a_0 is most broadly applied, and is as follows [2]:

$$a_r/b_r = 1 - 0.091a_0 \tag{3.4}$$

When a raindrop falls, its axis of symmetry deviates from the customary vertical direction attributable to various aerodynamic effects, thereby establishing an inclination angle of the raindrop, which is a three-dimensional space. The orientation of a raindrop in space is shown in Fig. 3.2.

Assuming that the propagation path is horizontal, HV plane in the figure is the plane perpendicular to the propagation direction, yZ plane is the one containing the axis of symmetry of the raindrop in the propagation direction, while the dotted line is the axis of symmetry. γ (gamma) is the angle of raindrop's axis of symmetry to its projection in HV plane. σ is the angle between the projection of raindrop's axis of symmetry in HV plane and V vertical axis, namely, the inclination angle of the raindrop. σ is regarded as the most instrumental parameter when considering the propagation characteristics of a raindrop. In general, assume σ approximately abides

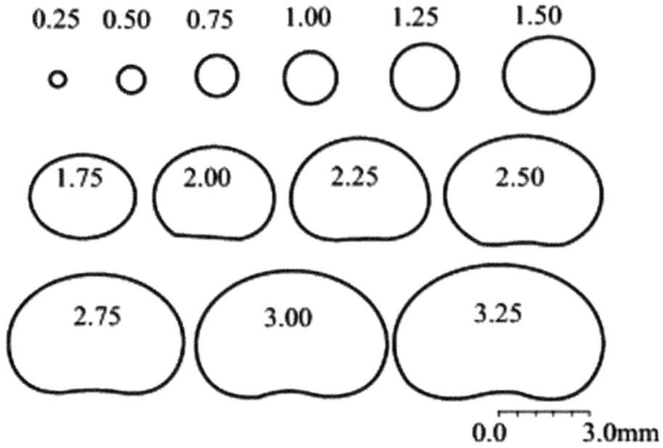

Fig. 3.1 Size and shape of a raindrop (numbers in the figure represent radius of equi-volume sphere) [2]

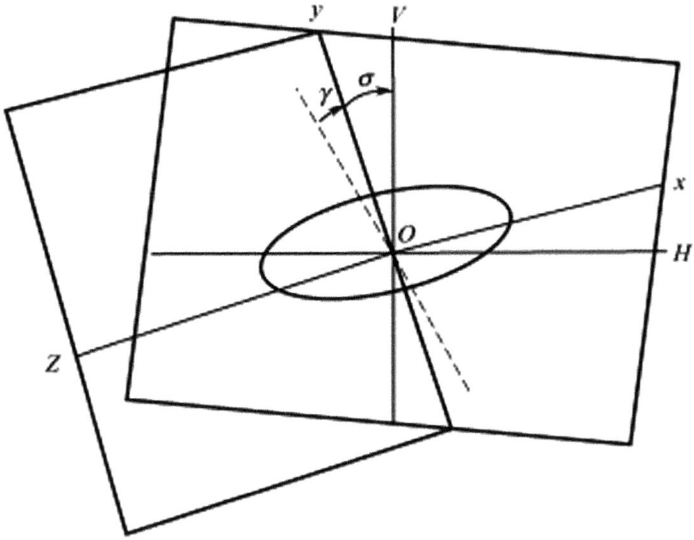

Fig. 3.2 Distribution of spatial inclination angle of raindrop

by Gaussian distribution:

$$p(x) = \frac{1}{\sqrt{2\pi}\sigma_0} e^{-\frac{(x-\mu_\sigma)^2}{2\sigma_0^2}} \tag{3.5}$$

In the formula: μ_σ is the average inclination angle of the raindrop; σ_0 is the standard deviation of the inclination angle. According to the long-term measurement of inclination angle, the average value of inclination angle $\mu_\sigma = 9°$, and standard deviation $\sigma_0 = 36°$.

End speed of raindrop remains an essential parameter for calculating the rate of rainfall through raindrop size distribution. Gunn, Kinder, and Best measured the end speed at an atmospheric pressure of 1013 hPa, relative humidity of 50%, and temperature of 20 °C. The measurement results illustrate that the falling speed of a raindrop increases with the increment in raindrop size. When raindrop diameter exceeds 2 mm, the increasing rate of its end speed gradually reduces. When raindrop diameter is 5 mm, its end speed is up to 9 m/s (max value), as shown in Table 3.2.

The end speed of the raindrop can be expressed by an empirical formula. Here, the empirical formula has been derived for a variation of end speeds. The following empirical formula has been fitted by Zhao Zhenwei in accordance with data in Table 3.2 [4]:

$$V(D) = 9.336 - \frac{24.065}{1 + 1.5301\exp(0.8165D)} \tag{3.6}$$

In the formula: $V(D)$ is the end speed of the raindrop (m/s); D is the equivalent diameter of the raindrop (mm).

Table 3.2 End speed of a raindrop [3]

Raindrop diameter D = 88	End speed of raindrop V/m/s	Raindrop diameter D/mm	End speed of raindrop V/m/s
0.1	0.256	2.8	7.82
0.2	0.72	3.0	8.06
0.3	1.17	3.2	8.26/
0.4	1.62	3.4	8.44
0.5	2.06	3.6	8.60
0./	2.47	3.6	8.72
0.7	2.67	4.0	8.83
0.6	3.27	4.2	8.92
0.9	3.67	4.4	8.98
1.0	4.03	4.6	9.03
1.2	4.64	4.6	9.07
1.4	5.17	5.0	9.09
1.6	5.65	5.2	9.12
1.6	6.09	5.6	9.14
2.0	6.49	5.8	9.1/
2.2	6.90	6.0	9.17
2.4	7.27	6.5	9.17
2.6	7.57	7.0	9.17

2. Classical drop size distribution:

The size distribution of a raindrop refers to its spatial distribution conditions with respect to different rainfall rates and sizes, which is also known as raindrop spectrum. Laws-Parsons raindrop size distribution (L-P distribution) measured by Laws and Parsons is a broadly adopted raindrop size distribution model [4]. Rain attenuation forecasting model of ITU-R is obtained by analyzing L-P distribution. L-P distribution is a discrete raindrop size distribution model, which registers the percentage of water content in an interval of 0.25 mm diameter each, to total water content under different rates of rainfall, as shown in Table 3.3.

L-P distribution provides raindrop diameter D (mm). The calculation formula for the number of raindrops within a unit space volume in diameter interval dD (mm) is given as follows [4]:

$$N(D)dD = \frac{10^2}{6\pi}\frac{R \cdot m(D) \cdot dD}{D^3 \cdot V(D)} \quad (\mathrm{m}^3) \tag{3.7}$$

In the formula: $m(D)$ denotes the raindrop volume distribution (m^{-3}), which can be checked in Table 3.3; $V(D)$ is the end speed of raindrop with diameter D (m/s); R is the rate of rain (mm/h).

In addition to L-P distribution, a myriad of researches have been conducted on size distribution and modeling using the observation data from different regions in the world, to generate a series of size distribution models, including negative exponential distribution, Gamma distribution, and logarithmic normal distribution, among which, negative exponential distribution is the most extensively used model, also called Marshall-Palmer distribution (M-P distribution) [5]:

$$N(D) = N_0 \exp(-f(R) \cdot D) \tag{3.8}$$

In the formula: N_0 is a constant and is also referred to as a function of rate of rainfall by certain researchers; $f(R)$ is a function of rate of rainfall.

The Joss model divides rainfall into drizzle, widespread showers, and thunderstorms and provides coefficients of negative exponential distribution under the three said conditions, as shown in Table 3.4 [6].

In 1980s and 1990s, Qiu Shengbo from China Research Institute of Radio Propagation measured raindrop spectrum of typical zones, such as temperate continental climate zone (Xinxiang), temperate marine climate zone (Qingdao), and subtropical marine climate zone (Guangzhou) [7]. Zhao zhenwei et al. evaluated the relationship between average raindrop size distribution and rate of rainfall in these three zones based on the measurement results. Overall, the average raindrop size distribution in China's different regions appeared identical to M-P and L-P distribution, as shown in Fig. 3.3, while the key difference lies in the content of small raindrops with a diameter less than 0.5 mm. According to the global measurement results, this difference seems quite prevalent and primarily caused by the diversity of different measurement

Table 3.3 Laws-Parsons raindrop size distribution [4]

Raindrop diameter/mm	Rate of rain/mm/h							
	Percentage of total volume/%							
	0.25	1.25	2.5	12.5	25	50	100	150
0.00–0.25	1.0	0.5	0.3	0.1	0	0	0	0
0.25–0.50	6.6	2.5	1.7	0.7	0.4	0.2	0.1	0.1
0.50–0.75	20.4	7.9	5.3	1.8	1.3	1.0	0.9	0.9
0.75–1.00	27.0	16.0	10.7	3.9	2.5	2.0	1.7	1.6
1.00–1.25	23.1	21.1	17.1	7.6	5.1	3.4	2.9	2.5
1.25–1.50	12.7	18.9	18.3	11.0	7.5	5.4	3.9	3.4
1.50–1.75	5.5	12.4	14.5	13.5	10.9	7.1	4.9	4.2
1.75–2.00	2.0	8.1	11.6	14.1	11.8	9.2	6.2	5.1
2.00–2.25	1.0	5.4	7.4	11.3	12.1	10.7	7.7	6.6
2.25–2.50	0.5	3.2	4.7	9.6	11.2	10.6	8.4	6.9
2.50–2.75	0.2	1.7	3.2	7.7	8.7	10.3	8.7	7.0
2.75–3.00	0.0	0.9	2.0	5.9	6.9	8.4	9.4	8.2
3.00–3.25	0.0	0.6	1.3	4.2	5.9	7.2	9.0	9.5
3.25–3.50	0.0	0.4	0.7	2.6	5.0	6.2	8.3	8.8
3.50–3.75	0.0	0.2	0.4	1.7	3.2	4.7	6.7	7.3
3.75–4.00	0.0	0.2	0.4	1.3	2.1	3.8	4.9	6.7
4.00–4.25	0.0	0.0	0.2	1.0	1.4	2.9	4.1	5.2
4.25–4.50	0.0	0.0	0.2	0.8	1.2	1.9	3.4	4.4
4.50–4.75	0.0	0.0	0.0	0.4	0.9	1.4	2.4	3.3
4.75–5.00	0.0	0.0	0.0	0.4	0.7	1.0	1.7	2.0
5.00 ~ 5.25	0.0	0.0	0.0	0.2	0.4	0.8	1.3	1.6
5.25–5.50	0.0	0.0	0.0	0.2	0.3	0.6	1.0	1.3
5.50–5.75	0.0	0.0	0.0	0.0	0.2	0.5	0.7	0.9
5.75–6.00	0.0	0.0	0.0	0.0	0.2	0.3	0.5	0.7
6.00–6.25	0.0	0.0	0.0	0.0	0.1	0.2	0.5	0.5
6.25–6.50	0.0	0.0	0.0	0.0	0.0	0.2	0.5	0.5
6.50–6.75	0.0	0.0	0.0	0.0	0.0	0.0	0.2	0.5
6.75–7.00	0.0	0.0	0.0	0.0	0.0	0.0	0.0	0.3

Table 3.4 Coefficients of negative exponential distribution [6]

Model	Name	N_0	$f(R)$
Marshall and Palmer	MP	8000	$4.1R^{-0.21}$
Joss-drizzle	JD	30,000	$5.7R^{-0.21}$
Joss-widespread	JW	7000	$4.1R^{-0.21}$
Joss-thunderstorm	JT	1400	$3.0R^{-0.21}$

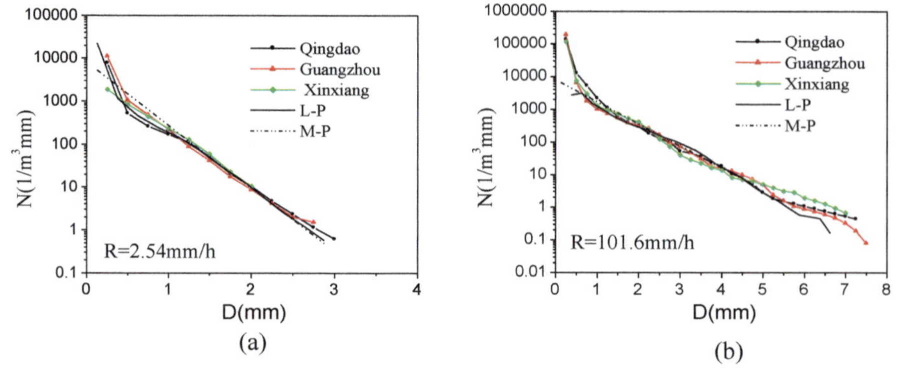

Fig. 3.3 Comparison between China's raindrop size distribution and L-P and M-P raindrop size distribution. **a** R = 2.54 mm/h; **b** R = 101.6 mm/h

and data processing methods. Currently, the content of small raindrops remains an ambiguous matter [8].

3. Spatial distribution of rainfall:

Rainfall along the tracks of links exhibits heavy variation when the satellite links pass through rain area. Figure 3.4 provides typical observation results of a rainfall event captured by a meteorological radar. According to the observation, there always exist one or more small areas displaying a high rate of rainfall interlined with vast zones showing a low rate of rainfall, known as the "rain cell." The size of the rain cell is relative to the rainfall rate of a given threshold value. Most of the time, the size of a rain cell is smaller than 10 km, and its size and intensity vary with respect to space and time due to the influence of wind. Rainfall rate also varies with altitudes together with horizontal direction. Altitude-varying rainfall rate is decided by zero-layer height, which regulates the melting process. Zero-layer height differs with seasons and zones. In general, the higher the average temperature of the earth's surface, the higher the zero layer. The zero-layer height decides the distance of satellite links passing through the rain area, which is instrumental in exploring the influence of satellite signal on spatial distribution.

According to the measured spatial distribution of rain rate, several rain cell models have been proposed at present. For example, cylindrical model, Gaussian model, EXCELL model [9], HYCELL model [9], and MultiEXCELL model, as shown in Fig. 3.5 These rain cell models are employed for statistical modeling and simulation of rain attenuation [10, 11].

4. Cumulative distribution of rainfall:

Long-term cumulative distribution of rainfall rate acts as a major requisite data required for statistical modeling of rain attenuation.

The cumulative distribution of rainfall rate is derived by making a statistical probability p $(R \geq R_i)$ of rainfall rate larger than and equal to setting value R_i.

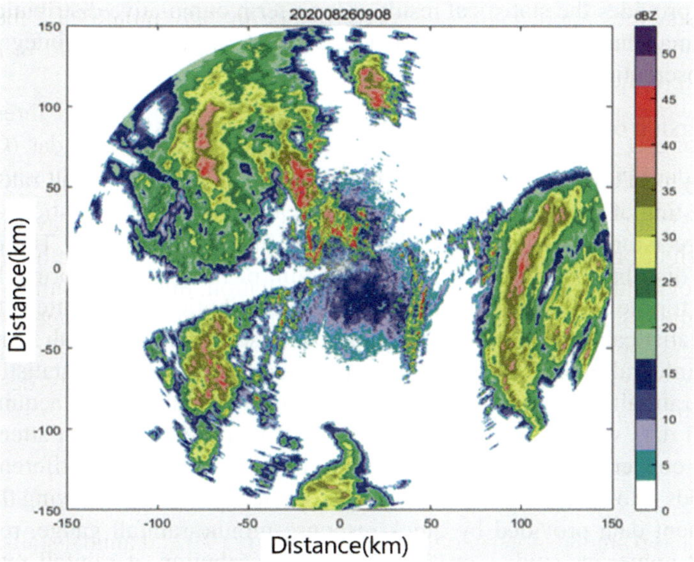

Fig. 3.4 Example of meteorological radar result

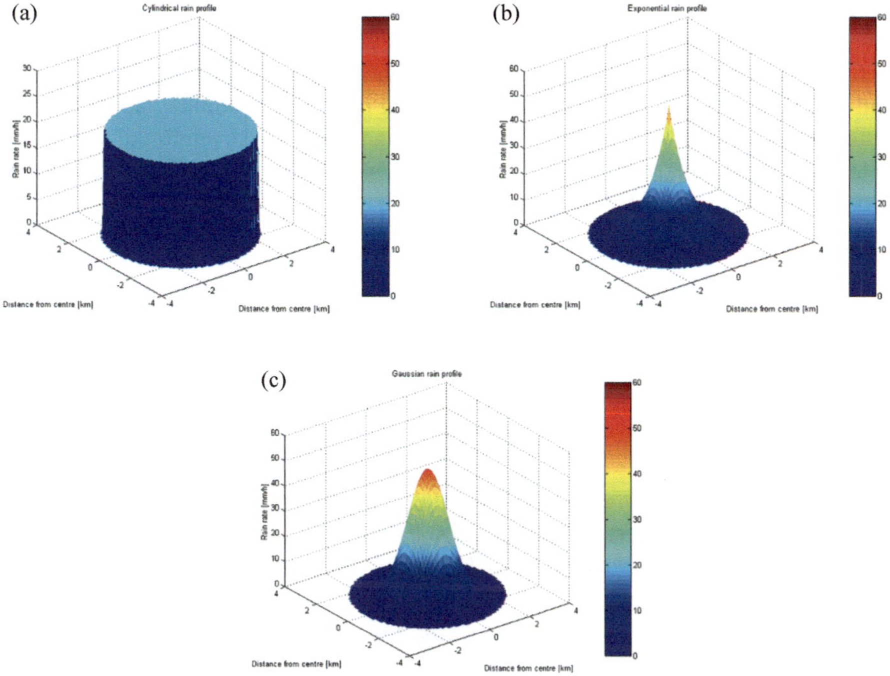

Fig. 3.5 Example of typical rain cell model. **a** Cylindrical rain cell; **b** Exponential rain cell; **c** Gaussian rain cell

Table 3.5 provides the statistical result of long-term cumulative distribution of rain rate in China's main cities with a statistical cycle of 10 years and an integral time of rainfall observation as 10 min.

5. Conversion of rainfall rates with different integral times:

According to ITU-R's recommendation, the statistical data of rainfall rate essential for forecasting of rain attenuation is primarily based on rainfall statistics with 1 min of time precision (usually being referred to as integral time) [12]. Therefore, the statistical distribution of rainfall rate with 1 min of integral time forms the source data used for the forecasting of the cumulative distribution of rain attenuation. The existing statistical distribution of rainfall rate is based on the data with 5 and 10 min or longer integral time, as observed in meteorological stations. The statistical distribution of rainfall rate for China's main cities listed in Table 3.5 is acquired based on rainfall data with 10 min of integral time. When predicting rain attenuation, it is vital to convert the statistical distribution of rainfall rate with different integral time periods into the distribution with 1 min of integral time. Employing the rainfall measurement data provided by quick-response minute-rainfall gauge, researchers conducted numerous studies on the statistical distribution of rainfall rate with 1, 5, 10 min, and longer integral time in different regions; and built the conversion model between the statistical distribution with different integral time and the statistical distribution with 1 min of integral time. Qiu Shengbo and others carried out an investigation on the conversion of rainfall rate with different integral time periods in 6 typical regions, including Nanjing, Qingdao, and Haikou [13]. Zhao Zhenwei and Lin et al. devised unified conversion model parameters, suitably applicable for different regions in Qiu's research. By combining the test data from Brazil and Korea, ITU derived a conversion model of rainfall rate with different integral time periods across the globe [14, 15]:

$$\mathrm{F}R_1(p) = a\left[R_\tau(p)\right]^b \tag{3.9}$$

In the formula: $R_1(p)$ and $R_r(p)$ represent rainfall rates (mm/h) with $p\%$ time probability exceeded, when the integral time is 1 min and Tmin, respectively; coefficient a and b are provided in Table 3.6.

6. Dielectric property of water:

The impact of hydrometeors on radio wave propagation is predominantly caused by the absorption and scattering of liquid water present in hydrometeors toward radio waves, which is a feature analogous to the complex dielectric characteristic of water. The complex dielectric characteristic of liquid water is always expressed by relative complex dielectric permittivity ($\varepsilon = \varepsilon' - i\varepsilon''$) or refractive index ($m = m' - im''$), and the relationship between the two is as shown below [16]:

$$m = \sqrt{\varepsilon} \tag{3.10}$$

Table 3.5 Cumulative distribution of rainfall rate in China's main cities

SN	City	Latitude	Longitude	Time probability/%									
				0.001	0.003	0.006	0.01	0.03	0.06	0.1	0.3	0.6	1
				Rainfall rate (mm/h)									
1	Beijing	39°48′	116°28′	96	71	60	48	28	17	11	4.8	2.1	1.1
2	Haikou	20°02′	110°21′	120	108	100	87	62	54	35	13	6.6	3.2
3	Nanning	22°49′	108°21′	115	102	90	80	54	40	28	10.7	5.1	2.8
4	Guangzhou	23°08′	113°19′	120	106	100	80	60	42	32	13	7	4
5	Nanchang	28°40′	115°58′	100.5	87	75	64	42	30	21	10	5.1	3.1
6	Fuzhou	26°05′	119°17′	105	89	72	62	41	30	21	10	6.1	4.1
7	Jinan	36°21′	116°59′	101	90	75	60	32	20	12	5.1	2.3	1.3
8	Qingdao	36°09′	120°25′	100	80	67	59	32	21	13	5.9	2.3	1.3
9	Guiyang	26°35′	106°03′	110	82	69	60	37	23	17	8.5	4.5	2.6
10	Hangzhou	30°19′	120°12′	110	80	68	54	32	21	15	7.5	4.2	2.8
11	Hefei	31°51′	117°17′	110	90	70	60	32	21	13	6.6	3.2	2.1
12	Nanjing	32°00′	118°48′	110	90	70	55	31	21	13	6.5	3.1	2.0
13	Tianjin	39°06′	117°10′	110	90	68	54	30	17	11	4.3	2.1	1.2
14	Wuhan	30°31′	114°04′	102	82	67	55	32	21	15	7.3	4.0	2.4
15	Changsha	28°12′	113°04′	100	80	66	55	32	21	17	8	4	2.3
16	Chengdu	30°40′	104°40′	100	78	63	53	31	21	15	6.2	3.1	2
17	Shenyang	41°46′	123°26′	100	73	60	48	25	17	11	5.1	2.6	1.3
18	Shanghai	31°10′	121°26′	91	71	60	50	30	21	13	7.0	3.6	1.2
19	Kunming	25°01′	102°41′	82	67	56	48	30	21	16	7.8	3.1	2.6
20	Changchun	43°54′	125°13′	97	70	56	45	22	13	10	4.1	2.1	1.2

(continued)

Table 3.5 (continued)

SN	City	Latitude	Longitude	Time probability/%									
				0.001	0.003	0.006	0.01	0.03	0.06	0.1	0.3	0.6	1
				Rainfall rate (mm/h)									
21	Shijiazhuang	38°02′	114°26′	92	70	51	41	21	12	10	4.2	2.1	1.2
22	Harbin	45°41′	126°17′	79	55	42	32	17	11	8	3.3	1.8	1.1
23	Taiyuan	37°47′	112°13′	78	51	39	30	15	10	7	3.2	1.7	1.1
24	Hohhot	40°49′	111°41′	68	54	32	27	14	10	7.6	3.4	1.8	1.1
25	Xi'an	34°18′	108°56′	58	37	27	20	11	8.2	6.1	3.5	2.1	1.3
26	Lanzhou	36°03′	103°53′	51	32	22	18	10	7	5	2.3	1.3	1.1
27	Lhasa	29°42′	91°08′	39	26	21	17	11	8.2	6.5	3.4	2.1	1.2
28	Xining	36°35′	101°55′	43	30	21	17	10	6.7	5	2.2	1.2	0.5
29	Yinchuan	38°29′	106°13′	52	31	21	17	8.2	5	3.8	2	1.1	0.7
30	Urumchi	43°34′	876°06′	15	11	9.1	8	5.2	4.1	3.2	1.8	1	0.60

Table 3.6 Conversion model coefficient of rainfall rates with different integral time periods [15]

T	a	b
5 min	0.986	1.038
10 min	0.919	1.088
20 min	0.680	1.189
30 min	0.564	1.288

Complex dielectric permittivity of water shows the dielectric complexity of this liquid as a function of temperature and frequency. For calculating the scattering characteristics of a raindrop, the Ray empirical formula is the most widely applied, as shown below [16]:

$$\varepsilon' = \varepsilon_\infty + \frac{(\varepsilon_s - \varepsilon_\infty)[1 + (\frac{\lambda_s}{\lambda})^{1-\alpha}\sin(\frac{\alpha\pi}{2})]}{1 + 2(\frac{\lambda_s}{\lambda})^{1-\alpha}\sin(\frac{\alpha\pi}{2}) + (\frac{\lambda_s}{\lambda})^{2(1-\alpha)}} \tag{3.11}$$

$$\varepsilon'' = \frac{(\varepsilon_s - \varepsilon_\infty)(\frac{\lambda_s}{\lambda})^{1-\alpha}\cos(\frac{\alpha\pi}{2})}{1 + 2(\frac{\lambda_s}{\lambda})^{1-\alpha}\sin(\frac{\alpha\pi}{2}) + (\frac{\lambda_s}{\lambda})^{2(1-\alpha)}} + \frac{\sigma\lambda}{18.8496 \times 10^{10}} \tag{3.12}$$

where

$$\begin{aligned}
\sigma &= 12.5664 \times 10^8 \\
\varepsilon_s &= 78.54[1 - 4.579 \times 10^{-3}(t - 25) \\
&\quad + 1.19 \times 10^{-5}(t - 25)^2 - 2.8 \times 10^{-8}(t - 25)^3] \\
\varepsilon_\infty &= 5.27134 + 2.16474 \times 10^{-2}t - 1.31198 \times 10^{-3}t^2 \\
\alpha &= -\frac{16.8129}{t + 273} + 6.09265 \times 10^{-2} \\
\lambda_s &= 3.3836 \times 10^{-4}\exp\left(\frac{2513.98}{t + 273}\right)
\end{aligned}$$

In the formula: λ refers to wavelength (cm); t refers to temperature (°C). Ray empirical formula is applicable to a temperature range of −20 to 50 °C. When the wavelength is shorter than 0.1 cm, taking into account the influence of the absorption belt becomes essential.

Another calculation of popular dielectric constant is by Double-Debye formula [17, 18]:

$$\varepsilon' = \frac{\varepsilon_0 - \varepsilon_1}{[1 + (f/f_p)^2]} + \frac{\varepsilon_1 - \varepsilon_2}{[1 + (f/f_s)^2]} + \varepsilon_2 \tag{3.13}$$

$$\varepsilon'' = \frac{f(\varepsilon_0 - \varepsilon_1)}{f_p[1 + (f/f_p)^2]} + \frac{f(\varepsilon_1 - \varepsilon_2)}{f_s[1 + (f/f_s)^2]} \tag{3.14}$$

In which,

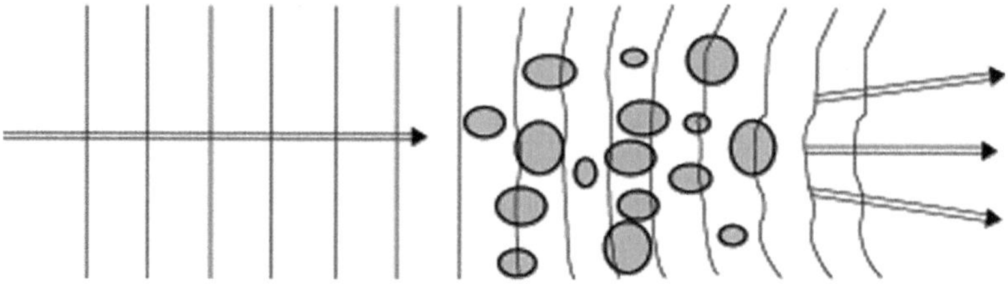

Fig. 3.6 Scattering sketch of hydrometeors

$$\varepsilon_0 = 77.66 + 103.3(\theta - 1)$$
$$\varepsilon_1 = 5.48$$
$$\varepsilon_2 = 3.51$$
$$f_p = 20.09 - 142.4(\theta - 1) + 294(\theta - 1)^2$$
$$f_s = 590 - 1500(\theta - 1)$$
$$\theta = 300/T$$

The frequency for Double-Debye formula ranges between 0 and 1,000 GHz.

Both the calculation methods for complex dielectric constant appear very consistent within 200 GHz. When the frequency is higher than 200 GHz, the calculation results differ to a certain extent. When calculating propagation features of rainfall, the Ray empirical formula is considered a standard approach, and when calculating propagation features of clouds and fog, the Double-Debye formula shows more prevalence.

7. Scattering characteristics of hydrometeors:

When incident waves are projected onto the scatterer, part of the incident energy is absorbed by the scatterer and converted into heat, and the other part gets scattered by it, as shown in Fig. 3.6.

The ratio of energy P_a absorbed by the scatterer to incident power density S_I (W/m^2) and of energy P_s scattered to incident power density S_I (W/m^2) are known as absorption section Q_a and scattering section Q_s, respectively; and the ratio of absorption section Q_a to physical section of the scatterer and of scattering section Q_s to physical section of the scatterer are called as normalized absorption cross section σ_a and normalized scattering cross-sectional σ_s, respectively. The sum of scattering cross section and absorption cross section is called as extinction cross-sectional Q_e. The sum of normalized absorption cross section and normalized scattering cross section is called as normalized extinction cross section σ_e and can be obtained using the following formulas [19]

$$Q_e = Q_a + Q_s \tag{3.15}$$

$$\sigma_e = \sigma_a + \sigma_s \tag{3.16}$$

When radio waves pass through the rain belts, attenuation differences in dissimilarly polarized waves and variation in polarization status are triggered due to the scattering and absorption of non-spherical raindrops. A geometric illustration of raindrop scattering is as shown in Fig. 3.7. Assume that E^i refers to electric field of incident waves, and the waves are shot to raindrops at an angle α, its polarization status is expressed by unit vector e, K_i refers to unit vector of incident direction, E^s refers to the electric field of scattering waves, and K_s refers to unit vector of scattering direction. Assume that the incident waves possess unit values, then the scattering field of a far-field region can be expressed as [20]:

$$\vec{E}^{s} = \frac{\vec{f}\left(\vec{K}_{i}, \vec{K}_{s}\right)}{r} e^{-jkr} \tag{3.17}$$

In the formula: k refers to constant of free-space propagation, r refers to distance from original point to observation point; $f(K_i, K_s)$ refers to the scattering amplitude vector.

For scattering of electrically large non-spherical raindrops, calculating a strict analytical solution is quite impracticable, and there exists no effective approximation; therefore, only the numerical method can be applied to derive an approximate solution. In recent decades, the scattering of non-spherical raindrops has garnered immense attention, with the scientific community exploring the subject with a multitude of studies, so as to develop a variety of numerical calculation methods, such as perturbation method and point match method. Once the scattering amplitude is obtained, the extinction section can be measured in accordance with the forward scattering theorem [21]:

$$Q_e = -(4\pi/k)\mathrm{Im}\left[\vec{e} \cdot \vec{f}(K_i, K_i)\right] \tag{3.18}$$

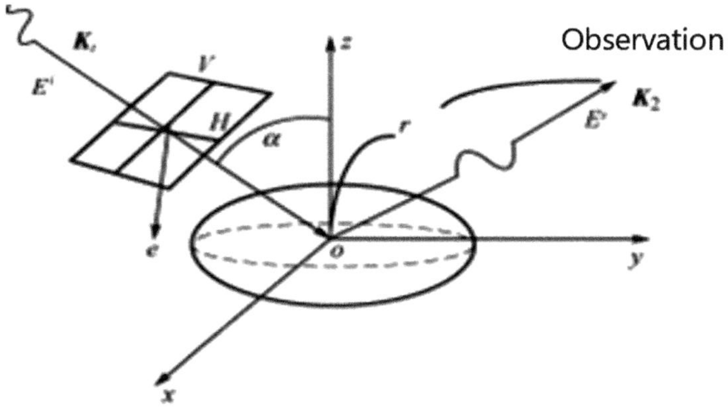

Fig. 3.7 Geometric representation of raindrop scattering

Without considering the polarization characteristics of radio waves, spherical raindrops can be used to replace common raindrops. Presently, the only viable and accurate solution of raindrop scattering can be derived through Mie scattering theorem, and the normalized scattering cross section and extinction cross section can be calculated by Mie coefficients [20].

$$\sigma_s(m, \chi) = \frac{2}{\chi^2}\sum_{n=1}^{\infty}(2n+1)(|a_n|^2 + |b_n|^2) \tag{3.19}$$

$$\sigma_e(m, \chi) = \frac{2}{\chi^2}\sum_{n=1}^{\infty}(2n+1)\mathrm{Re}(a_n + b_n) \tag{3.20}$$

In the formula: $\chi = (2\pi r)/\lambda_0$; $m = (m_r - jm_i) = \sqrt{\varepsilon_r - j\varepsilon_i}$ refers to refractive index; λ_0 refers to wavelength; r refers to media radius; a_n and b_n refer to Mie coefficients.

When the size of hydrometeors is relatively smaller than the wavelength of incident waves, that is, x $\ll$ 1, we can use Rayleigh scattering to approximately calculate the scattering characteristics of raindrops and fog drops, which is considerably easier than Mie theory and other numerical algorithms. Rayleigh scattering method is widely used for calculating the clouds-fog propagation characteristics and rainfall propagation characteristics at a low-frequency band. Its normalized scattering cross section and normalized absorption cross section are [21]:

$$\sigma_s = \frac{8}{3}\chi^4|K|^2 \tag{3.21}$$

$$\sigma_a = 4\chi\,\mathrm{Im}(-K) \tag{3.22}$$

Calculation formula of K is as shown below:

$$K = \frac{m^2 - 1}{m^2 + 1} = \frac{\varepsilon - 1}{\varepsilon + 2}.$$

The scattering and absorption cross section can be measured as follows [21]:

$$Q_s = \frac{2\lambda^2}{3\pi}\chi^6|K|^2 = \frac{128\pi^5 r^6}{3\lambda^4}|K|^2 \tag{3.23}$$

$$Q_a = \frac{\lambda^2}{\pi}\chi^3\mathrm{Im}(-K) = \frac{8\pi^2}{3\lambda}r^3\varepsilon_r\left|\frac{3}{\varepsilon + 2}\right|^2 \tag{3.24}$$

With Rayleigh approximation (x $\ll$ 1), the absorption cross section is much larger than the scattering cross section; therefore, the extinction cross section can be approximately expressed by using the absorption cross section.

The geometric relationship for scattering exhibited by the ellipsoid raindrops is as shown in Fig. 3.7. The scattering amplitude only consists of the component *e* when the incident electric field is in vertical polarization mode, that is, the polarization vector of the electric field *e* is in the plane *xOz*. Conversely, the scattering amplitude only entails the component *y* when the field is in horizontal polarization mode, that is, the polarization vector of the field *e* appears to be parallel to the *y*-axis. Let f_v and f_h represent the scattering amplitude in vertical and horizontal polarization mode, respectively; the amplitude can be calculated using the following equation for the oblate ellipsoid with short *a*-axis and long *b*-axis [21]:

$$f_v = k^2 \frac{ab^2}{3}\left[\frac{\varepsilon - 1}{1 + (\varepsilon - 1)A_1}\cos^2 a + \frac{\varepsilon - 1}{1 + (\varepsilon - 1)A_2}\sin^2 a\right] \tag{3.25}$$

$$f_h = k^2 \frac{ab^2}{3}\frac{\varepsilon - 1}{1 + (\varepsilon - 1)A_1} \tag{3.26}$$

In this equation, A_1 and A_2, being the geometric parameters of the ellipsoid, can be calculated by the following equation [21]:

$$A_1 = \frac{1}{2\left[\left(\frac{b}{a}\right)^2 - 1\right]}\left(\frac{\left(\frac{b}{a}\right)^2}{\sqrt{\left(\frac{b}{a}\right)^2 - 1}}\tan^{-1}\left(\sqrt{\left(\frac{b}{a}\right)^2 - 1}\right) - 1\right)$$
$$A_2 = 1 - 2A_1$$

The geometric parameters, A_1 and A_2, can be expressed as the following for the ellipsoid with long *a*-axis and short *b*-axis [21]:

$$A_1 = \frac{\frac{a}{b}}{2\left[\left(\frac{a}{b}\right)^2 - 1\right]}\left(\frac{a}{b} - \frac{1}{2\sqrt{\left(\frac{a}{b}\right)^2 - 1}}\ln\left(\frac{\frac{a}{b} + \sqrt{\left(\frac{a}{b}\right)^2 - 1}}{\frac{a}{b} - \sqrt{\left(\frac{a}{b}\right)^2 - 1}}\right)\right)$$
$$A_2 = 1 - 2A_1$$

The attenuation rate can be calculated by adopting the following equation collectively with the size distribution $n(r)$ of hydrometeors (rain droplets and cloud droplets) after deriving the extinction section [21]:

$$A = 4.343 \times 10^3 \int_0^\infty \sigma_e n(r)\, dr \quad \text{dB/km} \tag{3.27}$$

3.2 Cloud Attenuation

The extinction cross section of the cloud droplets in an air column can be calculated suitably by Rayleigh scattering approximation, since the size of the cloud droplet (1–60 μm) is smaller than the operating wavelength opted by the satellite communication system. In the Rayleigh approximation, the absorption cross section of the cloud droplets tends to be much larger than the scattering cross section, while the extinction section emerges approximately equal to the absorption section, so the extinction value can obtained by taking the sum of the absorption sections of all the cloud particles in a unit volume. The cloud attenuation rate I (dB/km) can be given as [19]:

$$I = 4.343 \times 10^3 \sum_{i=1}^{N} Q_a(r_i) \quad \text{dB/km} \tag{3.28}$$

where N is the number of particles per unit volume, and Q_a (r_i) is the absorption cross section of the particles with radius $r_{i.}$

The following equation can be obtained when combined with Eq. 3.24:

$$I = 4.343 \times 10^3 \times \frac{8\pi^2}{3\lambda}\varepsilon_r \left| \frac{3}{\varepsilon + 2} \right|^2 \sum_{i=1}^{N} r_i^3 \tag{3.29}$$

The water content W (g/m^3) is equivalent to the product of the total volume of cloud droplets per unit volume and density of water (10^6 g/m^3), that is

$$W = 10^6 \sum_{i=1}^{N} \frac{4\pi}{3} r_i^3 \tag{3.30}$$

Thus, the equation of cloud attenuation rate using the Rayleigh approximation can be given as follows [22]:

$$I = K_l W \tag{3.31}$$

$$K_l = \frac{0.819f}{\varepsilon''(1 + \eta^2)} \tag{3.32}$$

$$\eta = \frac{2 + \varepsilon'}{\varepsilon''} \tag{3.33}$$

where f is the frequency (GHz); K_l is the attenuation coefficient (dB/km)/(g/m^3); $\mathcal{E}'$ and $\mathcal{E}''$ are the real and imaginary parts of the complex dielectric constant of water, which can be calculated by Double-Debye Equation.

Temperature tends to be instrumental in affecting the process of cloud attenuation, since the complex dielectric constant of water behaves as a function of temperature.

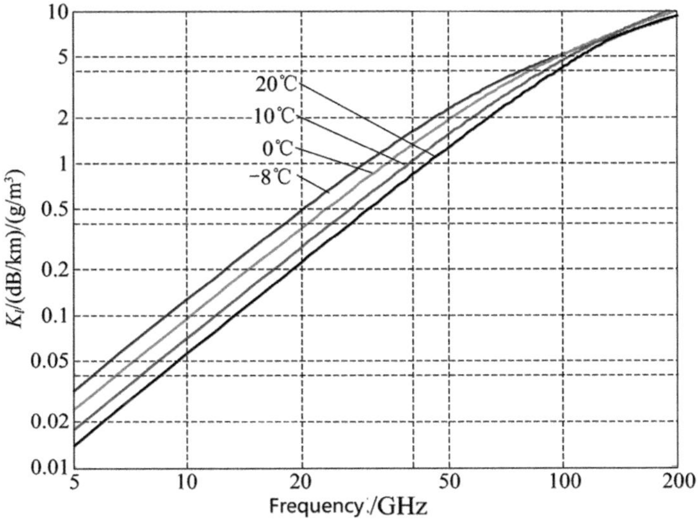

Fig. 3.8 Change in cloud attenuation coefficients at different temperatures with the frequency

The cloud attenuation coefficients within the frequency range of 5–200 GHz at −8, 0, 10, and 20 °C are shown in Fig. 3.8.

Zhao Zhenwei from the China Research Institute of Radio Propagation developed the empirical equation of K_l under the condition that Rayleigh be applied approximately, as the complex dielectric constant of water is the complex function of frequency and temperature makes the above-mentioned calculations for cloud attenuation unsuitable for engineering applications [23, 24].

When $f \leq 150$ GHz,

$$K_l = 6.0826 \times 10^{-4} f 1.8963 \left(\frac{300}{T}\right)^{(7.8087-0.01565f-3.0730\times10^{-4}f^2)} \tag{3.34}$$

When $150 < f \leq 1000$ GHz,

$$K_l = 0.07536 f^{0.9350} \left(\frac{300}{T}\right)^{(-0.7281-0.0018f-1.5420\times10^{-6}f^2)} \tag{3.35}$$

where f is the frequency (GHz); T is the temperature (K).

It has been verified that this empirical equation has a calculation error of less than 9% within the frequency range of 10–1000 GHz for temperatures between −8 and 20 °C.

While designing a satellite communication system, it is quite vital to consider the cloud attenuation rate exceeding at different time probability (p%) on the travel path, and likewise, it is necessary to obtain the different local integral water content $L(p)$ of cloud droplets exceeding at p% of the time (total mass of liquid water in the cylinder

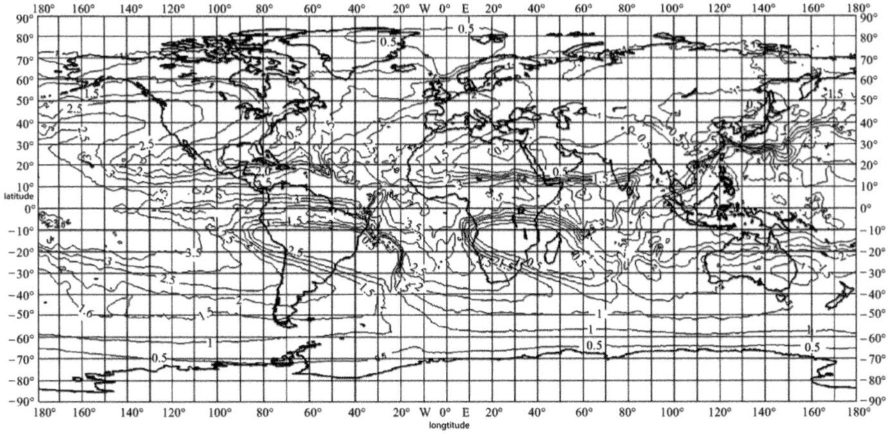

Fig. 3.9 Integral water content of cloud exceeded at 1% of the time (kg/m^2) [22]

with 1 m^2 of the area of base, kg/m^2). In case the data of the local integral water content of the clouds is unavailable or inaccessible, the digital map of liquid water content distribution in the clouds around the world, proposed in the ITU-R P.840 Proposal, can be employed. Figure 3.9 illustrates the integral water content exceeded at 1% of the time. The local integral water content in the clouds can be acquired through interpolation. Currently, we can use the following equation to predict the cloud attenuation $A(p)$ (dB) exceeding at different $p\%$ of the time [22]:

$$A(p) = \frac{L(p)K_l}{\sin\theta} \tag{3.36}$$

where θ is the angle of elevation of the path (°). Equation 3.36 is applicable to the condition of $5^\circ \leq \theta \leq 90^\circ$.

The prediction method for fog attenuation tends to be analogous with the cloud attenuation. Typically, the water content in fog W (g/m^3) or visibility V (km) is used to characterize their density. The relationship between the water content in fog and visibility can be shown in the form of the following empirical equation [19]:

$$\text{Advection fog: } W = (18.35V)^{-1.43} = 0.0156V^{-1.43} \tag{3.37}$$

$$\text{Radiation fog: } W = (42.0V)^{-1.54} = 0.00316V^{-1.54} \tag{3.38}$$

Assuming the length of an electric wave through the fog area is L_f (km), the attenuation generated by the fog can be calculated via the following equation:

$$A = K_l L_f W \quad \text{dB} \tag{3.39}$$

Considering the satellite-ground link design of a satellite communication system, the effect produced on it by the fog attenuation can be omitted due to the low height of the fog belt and short length of the electric wave through the fog area, unless specified otherwise.

3.3 Rain Attenuation

Rain attenuation serves as a major influencing factor governing the performance of a satellite communication system using the Ku and above band frequency. The statistical distribution of rain attenuation functions as the decisive groundwork for the link stability and design of the system allowance. The prediction method for rain attenuation has been evolving and strengthening perpetually with the accumulation of increasing test data and comprehensive research, along with multiple prediction models devised through extensive research.

3.3.1 Rain Attenuation Rate

In addition to the frequency and rainfall rate, the rain attenuation rate is also associated with the signal polarization methods, rain drop shape, and size distribution. The rate can be calculated by combining the rain drop size distribution with a scattering algorithm. To cater to various engineering applications, several studies have been carried out on the relationship between the rain attenuation rate, rainfall rate, and rain drop size distribution. According to Olsen et al., a commendable exponential relationship exists between the rain attenuation rate and the rainfall rate, as shown below [25]:

$$\gamma_R = kR^{\alpha} \tag{3.40}$$

where γ_R is the rain attenuation rate in dB/km; k and α are the regression coefficients associated with the polarization of electric wave; R is the rainrate in mm/h.

Based on the analysis of L-P raindrop size distribution and scattering characteristics of the ellipsoidal rain droplets, the equation of k and α proposed in Recommendation ITU-R P.838 is as follows:

$$k = [k_H + k_V + (k_H - k_V)\cos^2\theta\cos 2\tau]/2 \tag{3.41}$$

$$\alpha = [k_H\alpha_H + k_V\alpha_V + (k_H\alpha_H - k_V\alpha_V)\cos^2\theta\cos 2\tau]/2k \tag{3.42}$$

where T is the polarization ellipse (0° for horizontal polarization, 90° for vertical polarization, and 45° for circular polarization); coefficient k_H and k_V are calculated using Eq. 3.43, and coefficient α_H and α_V using Eq. 3.44

$$\log_{10} k_{H,V} = \sum_{j=1}^{3}\left(a_j \cdot \exp\left[-\left(\frac{\log(f) - b_j}{c_j}\right)^2\right]\right) + m_k \cdot \log(f) + c_k \quad (3.43)$$

$$\alpha_{H,V} = \sum_{i=1}^{4}\left(a_i \cdot \exp\left[-\left(\frac{\log(f) - b_i}{c_i}\right)^2\right]\right) + m_\alpha \cdot \log(f) + c_\alpha \quad (3.44)$$

Relevant parameters are shown in Tables 3.7, 3.8, 3.9 and 3.10.

Here, we try to precisely understand the change of relevant parameters during the calculation of rain attenuation rate with frequency. Figures 3.10 and 3.11 illustrate the horizontal polarization coefficient k_H and α_H, and the change curve of vertical polarization coefficient k_V and α_V with frequency.

Table 3.7 Parameters for k_H calculation [26]

j	a_j	b_j	c_j	m_k	c_k
1	–5.33980	–0.10008	1.13098	–0.18961	0.71147
2	–0.35351	1.26970	0.45400		
3	–0.23789	0.86036	0.15354		
4	–0.94158	0.64552	0.16817		

Table 3.8 Parameters for k_V calculation [26]

j	a_j	b_j	c_j	m_k	c_k
1	–3.80595	0.56934	0.81061	–0.16398	0.63297
2	–3.44965	–0.22911	0.51059		
3	–0.39902	0.73042	0.11899		
4	0.50167	1.07319	0.27195		

Table 3.9 Parameters for α_H calculation [26]

j	a_j	b_j	c_j	m_α	c_α
1	–0.14318	1.82442	–0.55187	0.67849	–1.95537
2	0.29591	0.77564	0.19822		
3	0.32177	0.63773	0.13164		
4	–5.37610	–0.96230	1.47828		
5	16.1721	–3.29980	3.43990		

Table 3.10 Parameters for α_V calculation [26]

j	a_j	b_j	c_j	m_α	c_α
1	–0.07771	2.33840	–0.76284	–0.053739	0.83433
2	0.56727	0.95545	0.54039		
3	–0.20238	1.14520	0.26809		
4	–48.2991	0.791669	0.116226		
5	48.5833	0.791459	0.116479		

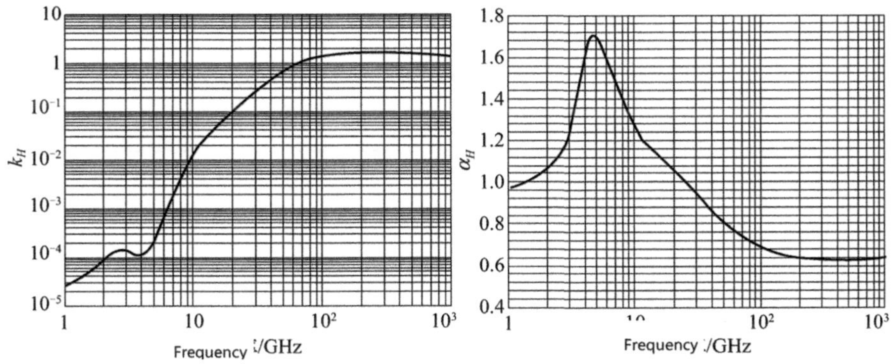

Fig. 3.10 Frequency relationship between k_H and α_H [25]

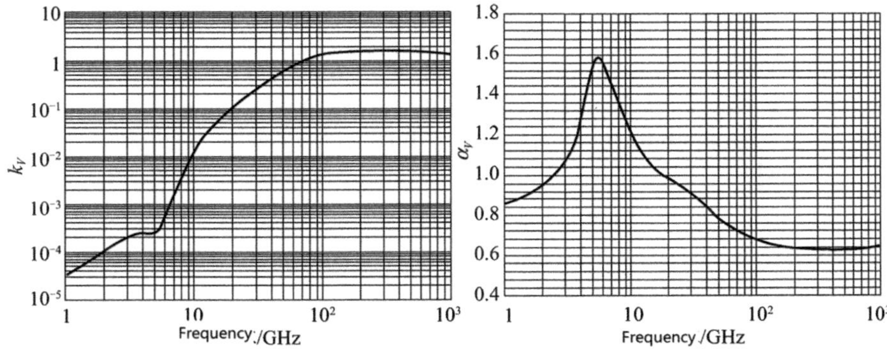

Fig. 3.11 Frequency relationship between k_V and α_V [25]

3.3.2 ITU-R Rain Attenuation Prediction Method [27]

The prediction method proposed by ITU for computing the rain attenuation rate of ground-to-air link is widely used around the world. This method, which was proposed by Dissanayake et al., takes into account the non-uniformity of the rain on horizontal and vertical paths. It has proven to be quite an effective method for the

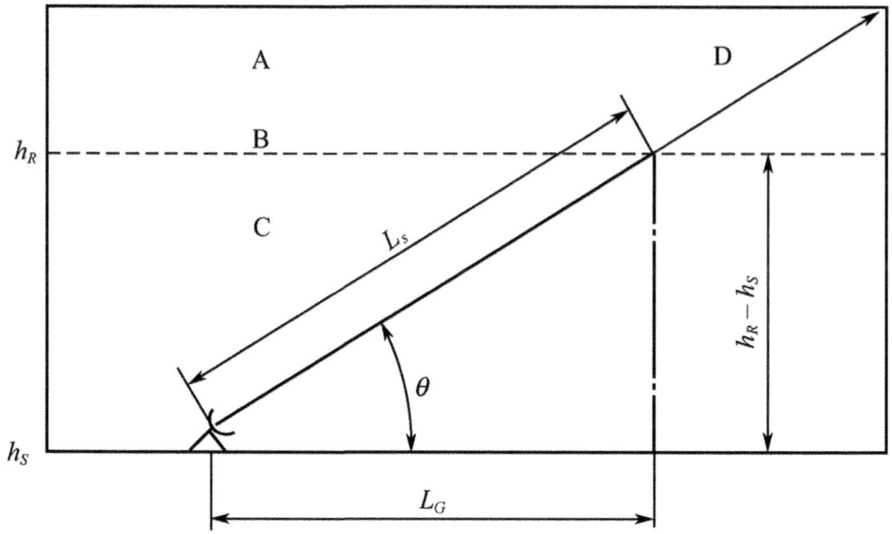

Fig. 3.12 Ground-to-air propagation path. (Area A: Frozen hydrometeor area; B: Rain height; Area C: Rainfall area; Area D: Ground-to-air path)

rain attenuation exceeded at 0.01% of the time probability, and several other methods have been derived from it.

The geometric representation of the electric wave propagation of ground-to-air link at the time of rain is as shown in Fig. 3.12. θ denotes the ground-to-air link communication elevation (°), h_s is the elevation of ground station (km), and h_R is the rain roof height (km), which can be calculated with the isothermal layer height h_0 at 0 °C (km); the equation for h_R proposed in ITU-R P.839 Proposal is $h_R = h_0 + 0.36$. L_s is the length of oblique propagation path of the electric wave through the raining area (km), which is calculated using the following equation:

$$L_s = \begin{cases} \frac{h_R - h_s}{\sin\theta} & \theta \geq 5^\circ \\ \frac{2(h_R - h_s)}{\sqrt{\sin^2\theta + \frac{2(h_R - h_s)}{R_e}} + \sin\theta} & \theta < 5^\circ \end{cases} \quad (3.45)$$

Re is the effective radius of the Earth. It is about 4/3 that of the actual radius of earth, which corresponds to approximately 8,500 km.

It can be noted that the rain attenuation rate is zero when the rain height h_R is less than the elevation h_s. L_G is the projection of L_s on the horizontal path (km): $L_G = L_s \cos\theta$.

The rain attenuation exceeded at 0.01% of the time $A_{0.01}$ (dB) of ground-to-air propagation link is calculated using the following equation:

$$A_{0.01} = \gamma_{R_{0.01}} \cdot L_E \quad (3.46)$$

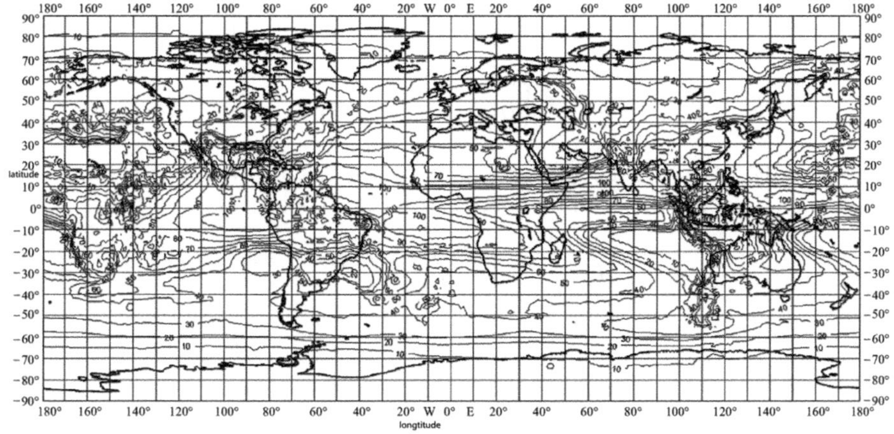

Fig. 3.13 Distribution map of annual rainfall rate exceeded 0.01% of the time

where $\gamma_{R0.01}$ is the rain attenuation rate (dB/km) with respect to the rainfall rate exceeded at 0.01% of the time $R_{0.01}$ (mm/h), which can be calculated using Eq. 3.40; $R_{0.01}$ is the local rainfall rate, with 1 min of integral time.

Proposal TULRP.837 recommended the digital map of the global statistical distribution of rainfall rate, shown in Fig. 3.13, which can be used as a reference when the local data remains unavailable or inaccessible. The rain attenuation rate is not required in case of $R_{0.01} = 0$.

In Eq. 3.46, L_E is the effective path length (km) with consideration of the irregularity of the rainfall (km), which can be calculated using the following equation:

$$L_E = L_R \upsilon_{0.01} \tag{3.47}$$

where

$$L_R = \begin{cases} L_R = \frac{L_G r_{0.01}}{\cos\theta} & \arctan\left(\frac{h_R - h_s}{L_G\, r_{0.01}}\right) > \theta \\ \frac{(h_R - h_s)}{\sin\theta} & \arctan\left(\frac{h_R - h_s}{L_G\, r_{0.01}}\right) \le \theta \end{cases} \tag{3.48}$$

$r_{0.01}$ is the horizontal adjustment factor, which can be defined as

$$r_{0.01} = \frac{1}{1 + 0.78\sqrt{\frac{L_G \gamma_R}{f}} - 0.38(1 - \exp(-2L_G))} \tag{3.49}$$

$u_{0.01}$ is the vertical adjustment factor, which can be defined as

$$\upsilon_{0.01} = \frac{1}{1+\sqrt{\sin\theta}\left(31\left(1-e^{-(\theta/(1+\chi))}\right)\frac{\sqrt{L_R\gamma_R}}{f^2}-0.45\right)}$$
$$\chi = \begin{cases} 36 - |\varphi| & |\varphi| < 36^\circ \\ 0 & |\varphi| \geq 36^\circ \end{cases} \tag{3.50}$$

where φ is the latitude of the ground station (°). The calculation equation for the rain attenuation exceeded at $p\%$ of the time A_p (dB) can be given as

$$A_p = A_{0.01}\left(\frac{p}{0.01}\right)^{-(0.655+0.033\ln(p)-0.045\ln(A_{0.01})-\beta(1-p)\sin\theta)} \tag{3.51}$$

$$\beta = \begin{cases} 0 & p \geq 1\%\text{或}|\varphi| \geq 36^\circ \\ \beta = -0.005(|\phi| - 36) & p < 1\%\text{且}|\varphi| < 36^\circ\text{且}\theta \geq 25^\circ \\ -0.005(|\varphi| - 36) + 1.8 - 4.25\sin\theta & \text{其他} \end{cases} \tag{3.52}$$

The annual average cumulative distribution of the rain attenuation can be measured through the ITU-R rain attenuation rate prediction method, however, the attenuation exhibits interannual variability due to the interannual variability of the rainfall. ITU-R model has been developed on the basis of the statistical data of ground-to-air rain attenuation and the corresponding rainfall rate. The majority of the test data is acquired from North America and Europe and minimally from low-latitude regions, hence, further testing is required for the suitable applicability of this model in the low-latitude regions. Meanwhile, the prediction result of ITU-R rain attenuation demonstrates a singularity variation with the elevation. Figure 3.14 illustrates the change of rain attenuation for the 30 GHz satellite-to-ground link for the ground station with the elevation. The propagation path through the rain area is shortened with the increase in elevation; however, the predicted rain attenuation remains raised. The credibility of such a prediction result has not been supported by the test data yet, which primarily influences the prediction result in the low-latitude regions with high satellite elevation.

3.3.3 Prediction Methods in China

To boost the accuracy and reliability of the forecasting of rain attenuation, full-probability rain attenuation forecasting methods are being developed around the world, which indicate the possibility of predicting other rain attenuations exceeded at X% of the time, in case of the data deficiency about the rain rate exceeded at 0.01% of the time, utilizing the rain attenuation rate at any X% of the time. Scholars namely Zhao Zhenwei, Lin Leke, and Lu Changsheng, from the China Research Institute of Radio Propagation, established a new full-probability rain attenuation prediction method by replacing the path adjustment factors by the rain rate adjustment factors based on the index rain cell model, which is referred as the Chinese method hereinafter [28]. This method has been submitted to the ITU by the relevant Chinese

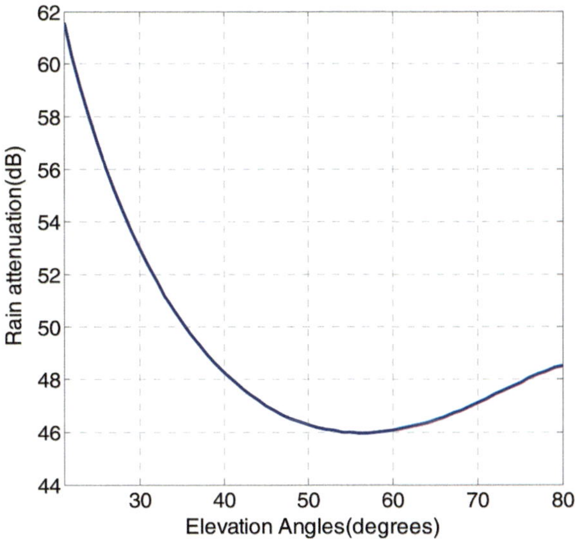

Fig. 3.14 Elevation singularity of ITU-R rain attenuation prediction method

authority. The Chinese method features better prediction accuracy and convenient processing, while also evading the singularity by comparing it to several prediction methods including ITU-R prediction.

In the Chinese method, the calculation equation for rain attenuation exceeded at $p\%$ of the time A_p (dB) is as follows:

$$A_p = k\left(r_p R_p\right)^{\alpha} L_s \tag{3.53}$$

where k and α are the coefficients for the calculation of rain attenuation (refer to Sect. 3.3.1 for calculation method); Rp is the rainfall rate (mm/h) exceeded at $p\%$ of the time; r_p is the rain rate adjustment factor. The equation is as follows:

$$r_P = 3.78 R_p^{-0.56+1.51/L_S}\left(1 - 0.85\frac{p^{0.065}}{1+0.12L_s}\right) \tag{3.54}$$

It should be noted that $r_p = 2.5$ in case of $r_p \geq 2.5$. Table 3.11 displays the comparison results of the prediction errors between the Chinese method and the available ITU method using the ITU-R satellite/ground link rain attenuation test database, which suggests a considerably higher accuracy of the Chinese method.

Table 3.11 Error comparison of rain attenuation prediction methods

Time probability	Prediction method	Average error %	RMS error %	Std error %
From 0.001 to 1%	ITU-R method	−5.31	31.78	31.33
	Chinese method	0.0025	24.41	24.41

3.3.4 Worst-Month Rain Attenuation

The performance standard of the satellite communication system is typically assessed in terms of the performance of "worst month" and "any month." The allowance design of the rain attenuation is generally determined by the rain attenuation in the worst months. Particularly, for the satellite broadcasting services (BBs), their performance is reviewed using monthly performance indicators. Therefore, the statistical characteristics of the worst-month rain attenuation are indeed crucial for the satellite communication system. The concept of worst months given by ITU-R is described as follows [29]:

(1) The time probability exceeding a given threshold in the worst months is referred to as the "annual time probability exceeding the threshold in the worst months."
(2) The statistics applicable to the performance in "any month" are the average long-term statistics of time probability exceeding the threshold in the worst month of the year.
(3) The worst month refers to the month that exceeds the given threshold for the longest time, which must be a natural month in the calendar (Gregorian calendar). The worst month may not be the same for different thresholds.

The model of rain attenuation prediction demonstrates the annual average cumulative distribution of rain attenuation. The rain attenuation in the months with heavy rainfall is evidently higher than the average attenuation; this may result in the availability of the system in the month much lower than the annual average availability, such that the application requirements are not fulfilled. The conversion equation of worst-month time probability and annual statistical time probability described in Recommendation ITU-RP.841 is as follows [26]:

$$p_w = Qp \tag{3.55}$$

where p_w and p are the worst-month time probability (%) and annual statistical time probability (%) under the same threshold, respectively; the equation of function Q is as follows [26]:

$$Q = \begin{cases} 12 & p < \left(\frac{Q_1}{12}\right)^{\frac{1}{\beta}} \\ Q_1 p^{-\beta} & \left(\frac{Q_1}{12}\right)^{\frac{1}{\beta}} < p < 3 \\ Q_1 3^{-\beta} & 3 < p < 30 \\ Q_1 3^{-\beta} \left(\frac{p}{30}\right)^{\frac{\lg(Q_1 3^{-\beta})}{\lg(0.3)}} & 30\% < p \end{cases} \tag{3.56}$$

The global conversion parameters $Q_1 = 2.85$ and $\beta = 0.13$ are described in Recommendation ITU-R P.841. Here, the relationship between Q and the annual probability p is as shown in Fig. 3.15, and the dotted line in the figure denotes the theoretical upper limit.

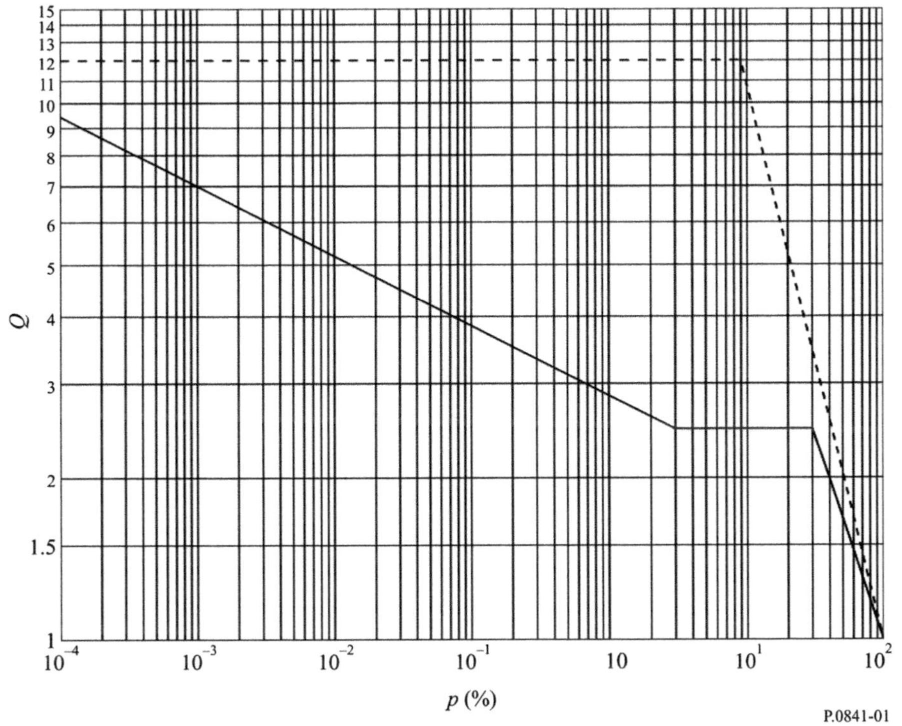

Fig. 3.15 Relationship between the global conversion parameter Q and p [30]

Zhan Zhenwei systematically evaluated the annual statistical distribution of global rainfall rate, including China, and statistical distribution characteristics in the worst month, and proposed two conversion methods of the rainfall rate in the worst month for two such regions categorized as—"tropical, subtropical, and temperate climate region with frequent rain" and "temperate, polar, and desert regions," which was incorporated in Recommendation ITU-R P.841. Table 3.12 displays the subregion parameters of the global rainfall rate in the worst month. Figure 3.16 illustrates the relationship between corresponding Q and annual average probability p.

To acquire a conversion method for the rainfall rate in the worst month that is more appropriate for China, we proposed three subregions (south, north, and arid region) in China based on the characteristics of annual average statistical distribution and worst-month statistic distribution, which corresponds to 1, 2, and 3 in Fig. 3.17,

Table 3.12 Subregion parameters of global rainfall rate in the worst month

Climate zone	Q_1	β
Tropical, subtropical, and temperate climate region with frequent rain	2.82	0.15
Temperate, polar, and desert regions	4.48	0.11

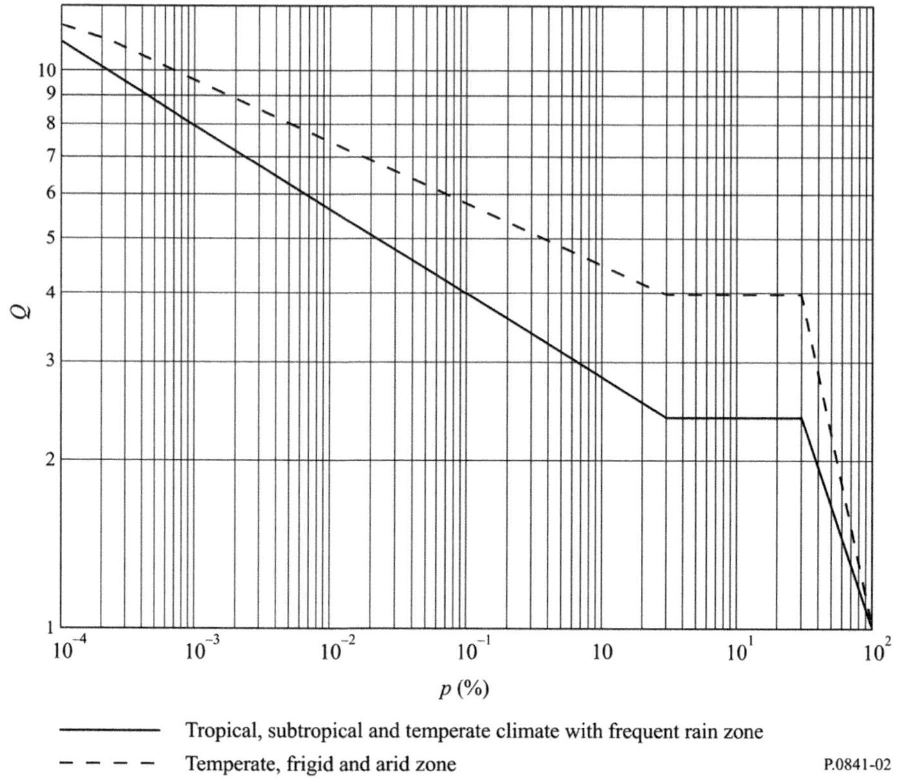

Fig. 3.16 Relationship between the subregion conversion parameters Q and p [26]

while also recommending the conversion parameters for these three subregions, as shown in Table 3.13.

For predicting the rain attenuation in the worst month, global conversion parameters can be utilized to obtain the worst-month statistical distribution by annual statistical distribution of the rain attenuation. Additionally, we can predict the worst-month rain attenuation in different regions of China by adopting the subregion parameters, since the rain attenuation is directly related to the rainfall characteristics.

3.4 Rain Attenuation Frequency Conversion

The rain attenuation frequency conversion refers to the measurement of the rain attenuation with a frequency, to predict the attenuation with another frequency. By and large, the rain attenuation with a high frequency is predicted with a low frequency. For satellite communication systems, we can calculate the rain attenuation for the

Fig. 3.17 Subregions of rainfall rate in the worst month

Table 3.13 Subregion parameters of rainfall rate in the worst month

Subregion	Q_1	β
South	3.12	0.15
North	4.12	0.13
Desert	5.40	0.10

downlink of the satellite by measuring the signal of the downlink, and then estimating the rain attenuation of the uplink by frequency conversion, to implement the anti-fading attenuation of the satellite system. In a rainfall event, the ratio of rain attenuation usually varies, increasing with a surge in rain attenuation. The frequency conversion of rain attenuation is divided into instantaneous frequency conversion and long-term statistic frequency conversion, which is introduced separately in the next sections.

3.4.1 Instantaneous Frequency Conversion

With the condition of measuring instantaneous rain attenuation A_1 (dB) of signal with frequency f_1, we can predict the possible distribution of rain attenuation A_2 (dB) of signal with frequency f_2 by instantaneous frequency conversion. Predominantly, this conversion method is utilized to control the uplink power and self-adaption coding modulation of the satellite communication system. This method is applicable for the rain attenuation with a frequency less than 55 GHz and requires cumulative distribution characteristics of the link rain attenuation with frequencies f_1 and f_2. According to the undertaken test, the cumulative distribution of rain attenuation for the satellite-to-ground link follows the logarithmic normal distribution, namely, $p(A_1 > a_1|A_1 > 0)$ and $p(A_2 > a_2|A_2 > 0)$ follow the logarithmic normal distribution of parameters (μ_1, σ_1) and (μ_2, σ_2), respectively [27].

$$p(A_1 > a_1 | A_1 > 0) = Q\left(\frac{\ln a_1 - \mu_1}{\sigma_1}\right) \quad (3.57)$$

$$p(A_2 > a_2 | A_2 > 0) = Q\left(\frac{\ln a_2 - \mu_2}{\sigma_2}\right) \quad (3.58)$$

where μ_1 and μ_2 are the average values of the logarithmic normal distribution, and σ_1 and σ_2 are the standard deviation values of the same; function Q is as follows [27]:

$$Q(x) = \frac{1}{\sqrt{2\pi}} \int_x^{\infty} \exp\left(-\frac{t^2}{2}\right) d_t \quad (3.59)$$

Here, the probability of uplink rain attenuation with frequency f_2 is shown when making $p(A_2 > a_2 | A_2 = a_1)$ represent the low-link rain attenuation $A_1 = a_1$ (dB) with f_1. Its prediction equation is as follows [27]:

$$p(A_2 > a_2 | A_1 = a_1) = Q\left(\frac{\ln(a_2) - \mu_{2/1}}{\sigma_{2/1}}\right) \quad (3.60)$$

When the rain attenuation with frequency f_1 is a_1, the prediction equation of instantaneous rain attenuation a_2 with frequency f_2 exceeded at the given time p is as follows [27]:

$$a_2 = \exp\left(\sigma_{2/1} Q^{-1}(p) + \mu_{2/1}\right) \quad (3.61)$$

where $\mu_{2/1}$ and $\sigma_{2/1}$ represent the conditional standard and conditional standard deviation, for which the equation is as follows [27]:

$$\mu_{2/1} = \frac{\sigma_2}{\sigma_1}\sqrt{1 - \xi^2} \ln(a_1) + \left(\mu_2 - \frac{\sigma_2 \mu_1}{\sigma_1}\sqrt{1 - \xi^2}\right) \quad (3.62)$$

$$\sigma_{2/1} = \sigma_2 \xi \quad (3.63)$$

where the length ξ is a coefficient related to frequency [27]:

$$\xi = 0.19\left[\frac{f_2}{f_1} - 1\right]^{0.57} \quad (3.64)$$

It is important to obtain parameters (μ_1 and σ_1) and (μ_2 and σ_2) for this method. We can obtain these parameters by fitting the statistical results of local rain attenuation or by fitting the calculated results from the above-mentioned prediction model.

Next, we introduce the fitting of μ_1 and σ_1 by taking the downlink as an example.

(1) First, we obtained the rain probability on the satellite-to-ground link p_{rain} (%), derived from statistics to historical rainfall data in local or Recommendation ITU-RP.837.
(2) Then, we got the time probability $p_{1,i}$ ($i = 1, 2, \ldots, 12$) of the downlink with frequency f_1 to be 0.01, 0.02, 0.03, 0.05, 0.1, 0.2, 0.3, 0.5, 1, 2, 3, and the statistical result of 5% rain attenuation $A_{1,i}$ (i = 1, 2, ..., 12), derived from the historical observation data or calculated by the above-mentioned prediction model.
(3) We obtained the conditional probability $p_{1,i}^c = \frac{p_{1,i}}{p_{\text{rain}}}(i = 1, 2, \ldots, 12)$ of the rain attenuation and then converted the probability and corresponding attenuation sequence $[p_{1,i}^c, A_{1,i}]$ to $[Q^{-1}(p_{1,i}^c), \ln A_{1,i}]$.
(4) We fitted the equation $\ln A_{1,i} = \sigma_1 Q^{-1}(p_{1,i}^c) + \mu_1$ by adopting minimum Least Square method, to obtain the logarithmic normal distribution parameters μ_1 and σ_1.

Taking a link as an example, with an uplink frequency of 30 GHz and downlink frequency of 20 GHz.

Figure 3.18 illustrates the distribution result (up link: $\mu_1 = 0.1837$, $\sigma_1 = 0.9411$; downlink: $\mu_2 = 1.0219$, $\sigma_2 = 0.8909$) of the conditional probability of the uplink attenuation predicted by the above-mentioned methods when the downlink attenuation is 12 dB.

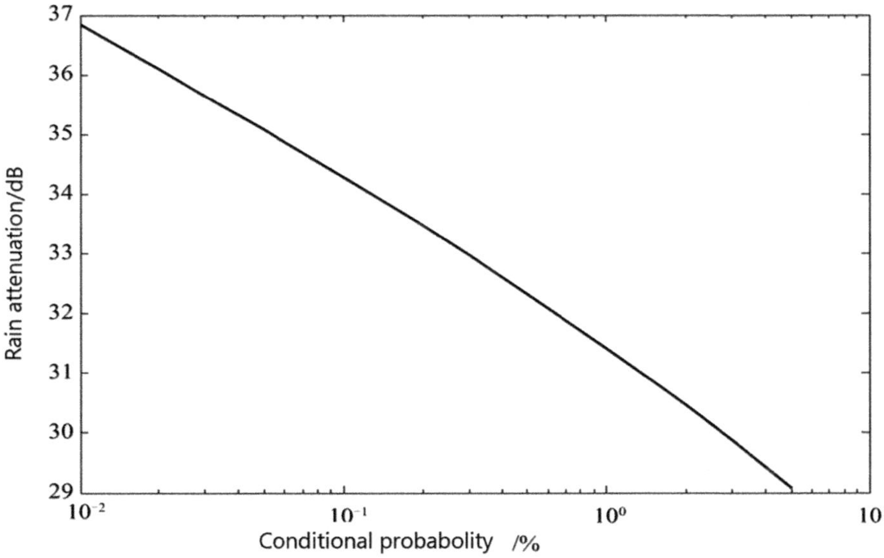

Fig. 3.18 Distribution result of the conditional probability of uplink attenuation

3.4.2 Long-Term Statistical Frequency Conversion of Rain Attenuation

When the statistical result of downlink attenuation is known, the result of the uplink attenuation can be easily predicted by the long-term statistical frequency conversion. This method is applicable to frequencies from 7 to 55 GHz. The equation of conversion from the uplink to the downlink is as follows [27]:

$$A_2 = A_1(\varphi_2/\varphi_1)^{1-H(\varphi_1,\varphi_2,A_1)} \tag{3.65}$$

where A_1 and A_2 are the equiprobable rain attenuation values with frequency f_1 (GHz) and f_2 (GHz), and:

$$\varphi_i = \frac{f_i^2}{1 + 10^{-4} f_i^2} \quad (i = 1, 2) \tag{3.66}$$

$$H(\varphi_1, \varphi_2, A_1) = 1.12 \times 10^{-3}(\varphi_2/\varphi_1)^{0.5}(\varphi_1 A_1)^{0.55} \tag{3.67}$$

While designing a satellite communication system, if the long-term measurement results of rain attenuation have been derived with a certain frequency for the same propagation path, then the prediction results of the attenuation with other frequencies for the same path can be obtained by utilizing the conversion method, which also presents a higher reliability quotient than the prediction model.

For pragmatic applications, the conversion calculation can be simplified by a fixed frequency conversion ratio.

3.5 Dynamic Characteristics of Rain Attenuation

With the advancement of satellite communication frequencies, statistical prediction of rain attenuation is a pre-requisite not only for system design, but also for the application and reliability evaluation of anti-fading techniques. The dynamic characteristics of the slant link rain attenuation are as shown in Fig. 3.19, including fading duration, fading interval, and fading slope. The fading duration is defined as the time interval between two intersections that exceed the threshold with the same attenuation. The fading interval is defined as the interval between two intersections that is lower than the threshold of the same attenuation. The fading slope is defined as the fading rate change with time. With respect to satellite communication, fading duration and fading slope are more important dynamic characteristics of rain attenuation. The prediction method for these two parameters is introduced in this section [30].

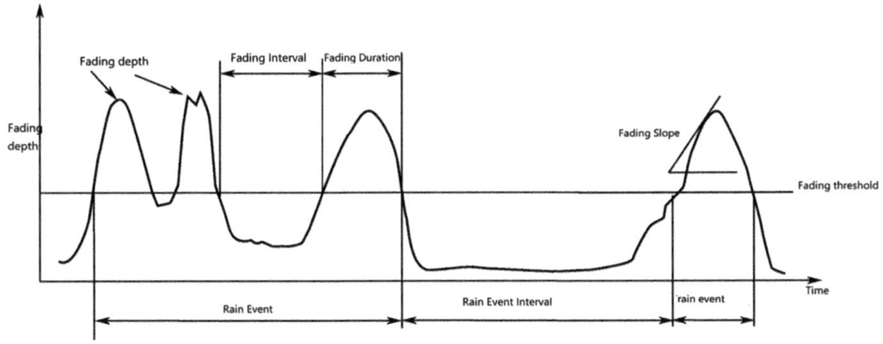

Fig. 3.19 Dynamic characteristics of rain attenuation of ground-to-air link

3.5.1 Fading Duration

The time and duration of the system interruption or failure caused by rainfall can be measured using the statistical characteristics of fading duration. The fading duration can be obtained from the resource allocation of the satellite system and can be used to provide the statistical duration of preserving the compensation state before the recovery of the anti-fading system to normal working mode. Furthermore, for satellite communication systems, the propagation channel usually generates an error code train instead of a single error code, and the characteristics of fading duration tend to support the coding pattern.

The fading duration is generally described in terms of two distributions: One is the occurrence probability of fading times p (Td > D|A > a), that is, the ratio of fading event times with duration $T_d > D(s)$ to all fading events exceeding the threshold with rain attenuation exceeding the threshold a (dB); and the second one is the cumulative exceeding probability of fading (Td > D|A > a), that is, the ratio of total fading time of fading duration $T_d > D(s)$ to total fading time exceeds the threshold with rain attenuation exceeding the threshold a (dB).

ITU-R adopts the two-sectional model to predict the fading duration, which entails logarithmic normal distribution of long-term fading and power exponential distribution; and can be suitably utilized for forecasting of fading duration for the satellite-ground links at frequency $10 \leq f \leq 50$ (GHz) and elevation angle ($5° \leq \theta \leq 60°$). When the given fading threshold is a (dB), the threshold duration of the boundary D_1 between long-term fading and short-term fading determines the following [30]:

$$D_t = D_0 \exp\left(p_1\sigma^2 + p_2\sigma - 0.39\right) \tag{3.68}$$

In which,

$$D_0 = 80\theta^{-0.4} f^{1.4} a^{-0.39} \tag{3.69}$$

$$\sigma = 1.85 f^{-0.05} a^{-0.027} \tag{3.70}$$

$$p_1 = 0.885\gamma - 0.814 \tag{3.71}$$

$$p_2 = -1.05\gamma^2 + 2.23\gamma - 1.61 \tag{3.72}$$

$$\gamma = 0.055 f^{0.65} a^{-0.003} \tag{3.73}$$

When fading duration is $1 \le D \le D_t$, it is regarded as short-term fading, and $p(T_d > D|A > a)$ and $F(T_d > D|A > a)$ are [30]:

$$p(T_d > D|A > a) = D^{-\gamma} \tag{3.74}$$

$$F(T_d > D|A > a) = \left[1 - k\left(\frac{D}{D_t}\right)^{1-\gamma}\right] \tag{3.75}$$

When fading duration is $D + D_t$, it is regarded as long-term fading, and the forecasting formulae of $p(T_d > D|A > a)$ and $F(T_d > D|A > a)$ are [30]:

$$p(T_d > D|A > a) = D_t^{-\gamma} \frac{Q\left(\frac{\ln(D) - \ln(D_2)}{\sigma}\right)}{Q\left(\frac{\ln(D_t) - \ln(D_2)}{\sigma}\right)} \tag{3.76}$$

$$F(T_d > D|A > a) = (1 - k)\frac{Q\left(\frac{\ln(D) - \ln(D_0)}{\sigma}\right)}{Q\left(\frac{\ln(D_t) - \ln(D_0)}{\sigma}\right)} \tag{3.77}$$

In which,

$$Q(z) = \frac{1}{\sqrt{2\pi}} \int_z^{\infty} e^{-\frac{1}{2}x^2} dx \tag{3.78}$$

$$k = \left[1 + \frac{\sqrt{D_0 D_2}(1 - \gamma) Q\left(\frac{\ln(D_t) - \ln(D_0)}{\sigma}\right)}{D_t \gamma Q\left(\frac{\ln(D_t) - \ln(D_2)}{\sigma}\right)}\right] \tag{3.79}$$

$$D_2 = D_0 \exp\left(-\sigma^2\right) \tag{3.80}$$

Moreover, when fading exceeds the threshold a, the forecasting formulae of $N(T_d > D|A > a)$ of fading duration time $T_d > D(s)$ and total fading time $T(T_d > D|A > a)$ are [30]:

$$N(T_d > D|A > a) = p(T_d > D|A > a) \times N_{\text{tot}}(a) \tag{3.81}$$

$$T(T_d > D|A > a) = F(T_d > D|A > a) \times T_{\text{tot}}(a) \tag{3.82}$$

In the formula: $N_{\text{tot}}(a)$ and $T_{\text{tot}}(a)$ represent the total number of attenuation events and total fading duration for which attenuation exceeds the threshold a, respectively, with the fading duration being longer than 1 s.

$T_{\text{tot}}(a)$ can be determined by locally measured data, or by the cumulative distribution of predicted rain attenuation, while $N_{\text{tot}}(a)$ can be calculated using the following formula [30]:

$$N_{\text{tot}}(A) = T_{\text{tot}}(A) \cdot \frac{k}{\gamma} \cdot \frac{1-\gamma}{D_t^{1-\gamma}} \tag{3.83}$$

The fading interval normally corresponds to the fading duration. Regarding the system functioning, it is highly crucial to identify statistical intervals after an interruption event and before the occurrence of the next one. According to test results, statistical characteristics of long-term fading interval appear in conformity with the logarithmic normal distribution, and short-term fading interval, similar to short-term fading duration, abides by the power exponential distribution, but a fully developed forecasting model and measurement method is yet to be formulated.

3.5.2 *Slope of Fading*

The slope of the fading parameter is utilized to design a control loop for tracking signal changes, as well as to predict short-term variation in propagation conditions, which is a key element for proper implementation of the satellite communication system in fade mitigation technology. The probability of slope of fading depends on various climatic parameters, raindrop size distribution, and rainfall type. For a given attenuation threshold, the slope of fading is inversely proportional to the path length and proportional to the elevation angle of the earth-space path. Furthermore, the slope of fading is closely related to the system parameters, and hence, receiving a window with a longer integral time will instantaneously decrease the change in fading. When analyzing the slope of fading of rain attenuation, it is essential to administer a low-pass filter for signals to eliminate any influence by troposphere scintillation. When forecasting the slope of fading, consideration of the parameters of the selected low-pass filter is equally significant.

Calculation formula of slope of fading $\zeta(t)$ (dB/s) at time t is as shown below [30]:

$$\zeta(t) = \frac{A\left(t+\frac{1}{2}\Delta t\right) - A\left(t-\frac{1}{2}\Delta t\right)}{\Delta t} \tag{3.84}$$

In the formula: A refers to rain attenuation (dB) at a proper moment; Δt refers to interval (s) for calculation of the slope of fading.

The forecasting method recommended in ITU-PR.1623 can be adopted for predicting slope of fading. With respect to rain attenuation value A, probability density function $p(\zeta|A)$ with ζ of slope of fading can be calculated using the following formula [30]:

$$p(\zeta|A) = \frac{2}{\pi\sigma_\zeta\left(1 + (\zeta/\sigma_\zeta)^2\right)^2} \tag{3.85}$$

Formula of complementary cumulative distribution $F(\zeta|A)$ with slope of fading exceeding ζ is [30]:

$$F(\zeta|A) = \frac{1}{2} - \frac{\zeta/\sigma_\zeta}{\pi\left(1 + (\zeta/\sigma_\zeta)^2\right)} - \frac{\arctan(\zeta/\sigma_\zeta)}{\pi} \tag{3.86}$$

Formula of conditional probability $F(\zeta|A)$ of absolute value with slope of fading exceeding ζ is [30]:

$$F(|\zeta||A) = 1 - \frac{2(|\zeta|/\sigma_\zeta)}{\pi\left(1 + (|\zeta|/\sigma_\zeta)^2\right)} - \frac{2\arctan(|\zeta|/\sigma_\zeta)}{\pi} \tag{3.87}$$

In which,

$$\sigma_\zeta = SF(f_B, \Delta t)A \quad \text{dB/s} \tag{3.88}$$

$$F(f_B, \Delta t) = \sqrt{\frac{2\pi^2}{\left(1/f_B^b + (2\Delta t)^b\right)^{1/b}}} \tag{3.89}$$

In the formula: $b = 2.3$ is a constant; s refers to parameters related to the climate and elevation angle. In Europe and United States, the total average value of elevation angle 10°–50° is 0.01; f_B refers to 3 dB cut-off frequency provided by the low-pass filter, using 0.001–1 Hz in general. The applicable ranges of this method are as follows: frequency $10 \leq f \leq 30$ (GHz), elevation angle of links $10° \leq \theta \leq 50°$, rain attenuation $0 < A \leq 20$ (dB), and interval of calculation of slope of fading $2 \leq \Delta t \leq 200$ (s).

Zhang et al., fitted the σ_ζ in the above-mentioned model once again, based on the rain attenuation observation data of 12.25 GHz satellite-ground link in China's Haikou, and derived a new calculation formula for σ_ζ [30]:

$$\sigma_\zeta = S(A) \times \sigma(\Delta t) \tag{3.90}$$

$$S(A) = -0.0128A^2 + 0.2874A - 0.3104 \quad (3.91)$$

$$\sigma(\Delta t) = -1.8106 \times 10^{-8}\Delta t^3 + 7.1973 \times 10^{-6}\Delta t^2 - 0.001\Delta t + 0.0839 \quad (3.92)$$

The coefficients in formula (3.91) and formula (3.92) are relative to the regions considered. By being tested by measured data, the upgraded model proves to possess a higher forecasting precision for Haikou than the ITU-R model.

3.6 Rain Attenuation Time Series

With the rapid advancement of frequency-based satellite communication, the link attenuation represented by rain attenuation has become increasingly acute. It is quite challenging to supplement propagation attenuation of links by preserving system allowance, since resources could be wasted and other systems strongly disturbed if a high allowance is planned. For optimization of system resources and enhancement of anti-fading ability of the system, satellite communication systems at the Ka frequency band and above are obligated to implement self-adaption anti-fading technologies, such as self-adaption power control, self-adaption modulation and coding, and self-adaption speed regulation. Obtaining the suitable time series of links for a rational design and effective implementation of the self-adaption anti-fading technologies is vital. With the limitations put forward by the test conditions, the measurement data of rain attenuation can only be obtained from a small number of stations, which creates a hindrance for the effective implementation of anti-fading design in a satellite system for a widespread area. The time series of any links can be simulated using the synthetic method of time series, so that this deficiency is supplemented very well.

The universal synthetic methods suggested in ITU-RP.1853 recommendation are as shown in Fig. 3.20. Following the synthesis process of the time series $A(t)$, a random white Gaussian noise $n(t)$ with mean value of 0 and standard deviation of 1 is filtered to generate a random sequence $X(t)$ and is then converted from the normal distribution to logarithmic normal distribution applying a memoryless non-linear component to synthesize a time series of links via offset calibration.

Determination of 5 statistical parameters is significant for synthesis, including rain probability p_{rain} (%), mean value μ, and standard deviation σ of logarithmic normal

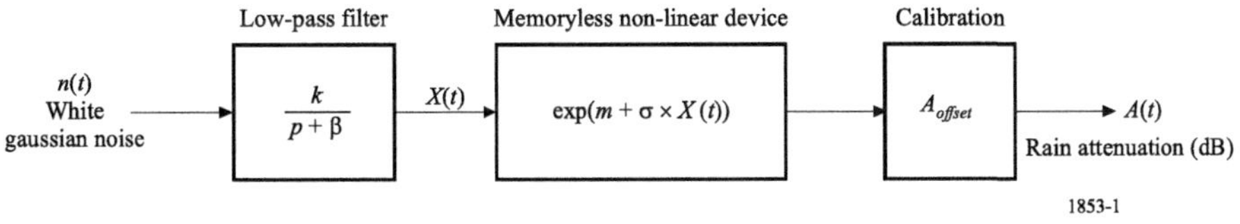

Fig. 3.20 Block diagram of the rain attenuation time series synthesizer [31]

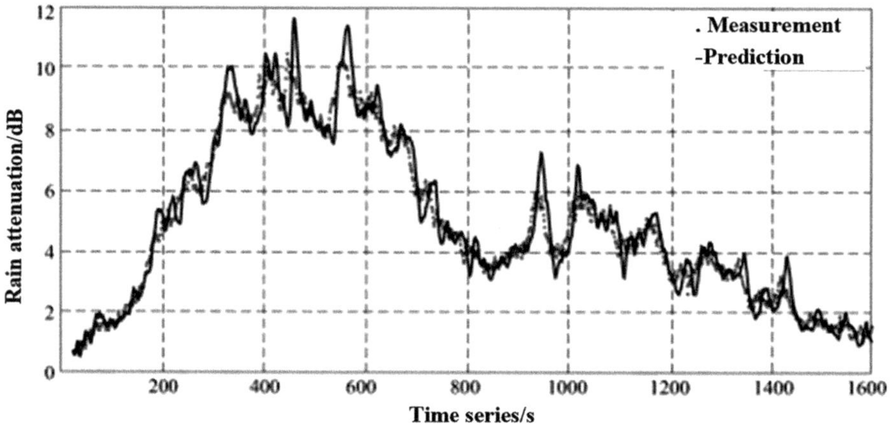

Fig. 3.22 Example of real-time forecasting based on gray theory

system resources can be executed in accordance with the forecasting result of rain attenuation region distribution. Figure 3.23 provides an example of forecasting the result of rain attenuation region.

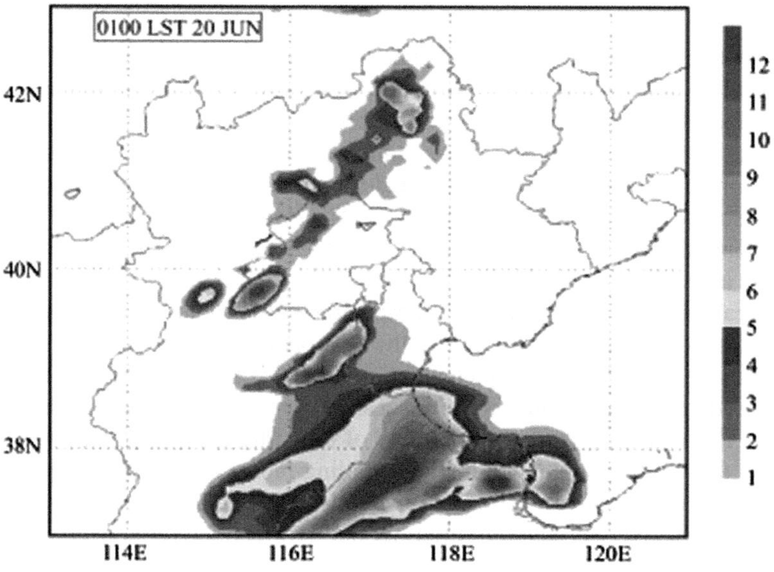

Fig. 3.23 Short-term forecasting of rain attenuation

3.8 Rain-Induced Cross-Polarization

When the radio waves of a satellite communication system travel along the slant path, the polarization direction of radio waves may get deflected, which is called as depolarization. With respect to satellite communication systems using orthogonal polarization multiplexing technology, the depolarization effects emerge as critical factors determining the performance of orthogonal channels. There exist diverse physical mechanisms of depolarization. Ground reflection and atmospheric stratification reflection in clear air can lead to depolarization, especially multipath propagation. During rainfall, the scattering of raindrops can lead to cross-polarization effect. With respect to satellite communication systems operating at 10 GHz and above, rain-induced depolarization is a key factor that causes depolarization of a satellite. This chapter mainly introduces the rain-induced depolarization effect and forecasting method of satellite signals.

3.8.1 Description of Depolarization

The original linearly and circularly polarized radio waves of a satellite become elliptically polarized under the depolarization effects of propagation media. Polarization characteristics of radio waves can be described using the following parameters:

① Rotation direction (looking in the opposite direction of the wave): Left rotation and right rotation.
② Axial ratio: Amplitude ratio of the long and short axis of the polarization ellipse.
③ Inclination angle of polarization ellipse: Inclination angle of polarization ellipse to coordinate satellite in the propagation direction.

By and large, cross-polarization discrimination (XPD) and cross-polarization isolation (XPI) are used to describe depolarization characteristics of radio waves. Take linear polarization as an example, as shown in Fig. 3.24, the XPD and XPI are defined as follows:

(1) Cross-polarization discrimination, noted as XPD. When transmitting a polarized radio wave, the ratio of co-polarization signal power and orthogonal polarization power obtained at the receiving terminal is defined as XPD (dB), that is

$$XPD = 20\lg\frac{E_{11}}{E_{12}} \quad \text{dB} \tag{3.97}$$

(2) Cross-polarization isolation, noted as XPI. The transmitting terminal transmits the radio waves with the same power in two orthogonal polarization directions, and the ratio of the power of co-polarized waves to the power of cross-polarized waves received in one polarization direction is defined as XPI (dB), that is

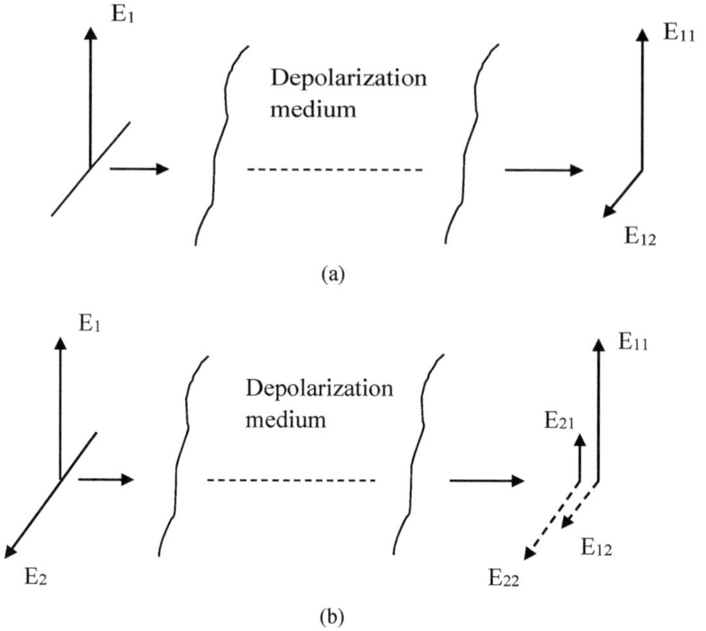

Fig. 3.24 Sketch of depolarization characteristics. **a** Cross-polarization discrimination (XPD); **b** Cross-polarization isolation (XPI)

$$XPI = 20\lg\frac{E_{11}}{E_{21}} \quad \text{Or} \quad 20\lg\frac{E_{22}}{E_{12}} \quad \text{dB} \tag{3.98}$$

When transmission signals have the same amplitude as the electric field ($E_1 = E_2$) and depolarization due to the receiving system can be overlooked, the measurement result of XPD is the same as that of XPI. Formulae (3.97) and (3.98) are applicable for linear polarization, but the definitions with respect to XPD and XPI are applicable for any other orthogonally polarized systems. The XPD is always used to measure polarization characteristics in earth-space path propagation.

3.8.2 Rain-Induced Depolarization Mechanism

The spherical raindrops do not cause depolarization effects, so long as the axis of raindrops coincides with the polarization electric field of vertically or horizontally polarized waves, even if the raindrops are ellipsoidal. When the axis of a raindrop is inclined against the wave polarization direction, there exists a difference in attenuation and phase offset between the two electric field components, as the path length that the vertical component and horizontal component of the electric field pass through is different, that is, there is differential attenuation and differential phase offset, such that the polarization status of outgoing waves is different from that of incident waves leading to cross-polarization. According to several studies, differential phase shift at a low-frequency band remains the principal cause of depolarization, but for frequencies between 30 and 90 GHz, it is the differential attenuation. Noticeably, rain-induced cross-polarization is caused by the offset of the raindrop axis and propagation direction and depends on the inclination angle of path and inclination angle of raindrop, among which, raindrop inclination angle is the most significant factor (Fig. 3.7). A raindrop's inclination angle is caused by air flow and is completely random. According to research, the inclination angle of raindrop abides by the Gaussian distribution, and its mean value and standard deviation are relative to the rainfall rate and vary with different regions. Oguchi suggests that inclination angles—σ and γ, of a raindrop are independent of the raindrop's size; they abide by Gaussian distribution and function independent of each other, and this is the most prevalent inclination angle distribution model used for the rain-induced cross-polarization algorithm.

3.8.3 Prediction Method of Rain-Induced Cross-Polarization

For rain-induced cross-polarization, extensive theories and test studies have been developed internationally. According to several studies, a good correlation is found between rain-induced cross-polarization discrimination and co-polarization attenuation. For the same time probability, there is a semi-empirical relationship between rain-induced cross-polarization and co-polarization attenuation [27]:

$$XPD = U - V\lg[A] \tag{3.99}$$

In the formula: *XPD* and *A* represent cross-polarization discrimination (dB) and co-polarization attenuation (dB), respectively, with the same being exceeded; *U* and *V* refer to undetermined parameters, and different prediction time probability methods are essentially reflected by different *U* and *V* values. In subsequent sections, we will introduce the rain-induced cross-polarization prediction method recommended by ITU (ITU-R prediction method) and an upgraded prediction method that is submitted to ITU by China (China's prediction method).

1. **ITU-R prediction method**:

A prediction method of rain-induced cross-polarization of earth-space links is provided in ITU-PR.618 Proposal, which is suitable for frequencies $6 \le f \le 55$ (GHz) and elevation angle of $\theta \le 60°$. Cross-polarization discrimination with p (%) time probability exceeded XPD_p (dB) is calculated using the following formula [27]:

$$XPD_p = XPD_{\text{rain}} - XPD_{\text{ice}} \tag{3.100}$$

In the formula: XPD is an ice-related item, $XPD_{ice} = XPD_{rain} \times (3.3 + 0.1 \lg p)/2$; XPD_{rain} is an item of rain-induced cross-polarization discrimination (dB), XPD_{rain} with p (%) time probability exceeded is calculated using the following formula [27]:

$$XPD_{rain} = C_f - C_A + C_\tau + C_\theta + C_\sigma \tag{3.101}$$

In which, frequency-related item, C_f (dB), is calculated using the following formula [27]:

$$C_f = \begin{cases} 60 \lg f - 28.3 & 6 \le f < 9\,\text{GHz} \\ 26 \lg f + 4.1 & 9 \le f < 36\,\text{GHz} \\ 35.9 \lg f - 11.3 & 36 \le f \le 55\,\text{GHz} \end{cases} \tag{3.102}$$

Rain attenuation-related item, C_A (dB), is calculated using the following formula [27]:

$$C_A = V(f) \lg A_p \tag{3.103}$$

$$V(f) = \begin{cases} 30.8 f^{-0.21} & 6 \le f < 9\,\text{GHz} \\ 12.8 f^{0.19} & 9 \le f < 20\,\text{GHz} \\ 22.6 & 20 \le f < 40\,\text{GHz} \\ 13.0 f^{0.15} & 40 \le f < 55\,\text{GHz} \end{cases} \tag{3.104}$$

In formula (3.103): A_p is rain attenuation with p (%) time probability exceeded, also called as co-polarization attenuation, and historically statistical results or model prediction results can be adopted.

C_τ (dB) is a polarization angle-related polarization improvement factor and is calculated using the following formula[27]:

$$C_\tau = -10 \lg[1 - 0.484(1 + \cos 4\tau)] \tag{3.105}$$

When $\tau = 0°$, $C_\tau = 0$; when $\tau = 45°$ or 90, C_τ is up to 15 dB.

C_θ (dB) is an elevation angle-related item and is calculated using the following formula [27]:

$$C_\theta = -40 \lg(\cos \theta) \tag{3.106}$$

C_σ (dB) is a raindrop inclination angle-related item and is calculated using the following formula [27]:

$$C_\theta = -40 \lg(\cos \theta) \tag{3.107}$$

In the formula: σ_0 is standard deviation of inclination angle distribution of raindrop (°). σ_0 values corresponding to 1%, 0.1%, 0.01%, and 0.001% time probability exceeded are 0°, 5°, 10°, and 15°.

2. **China's method**:

Zhao Zhenwei and Lin et al., using second-order small variable approximation of rain-induced cross-polarization theory, postulated an enhanced rain-induced cross-polarization forecasting method by fitting the parameters collectively with ITU's rain-induced cross-polarization data and submitted it to ITU for inspection, which, hereafter, is called as "China's method" [32]. Compared with ITU-R's method, a second-order parameter ΔXPD is added to China's method when calculating XPD_{rain}, to develop a new calculation formula for frequency-related item C_f and rain attenuation-related item C_A [32]:

$$XPD_{\text{rain}} = C_f - C_A + C_\tau + C_\theta + C_\sigma + \Delta XPD \tag{3.108}$$

$$C_f = 20 + 11\lg f \tag{3.109}$$

$$C_A = 21.7\lg A_p \tag{3.110}$$

$$\Delta XPD = -0.085A_p\cos^2\theta\cos 2\tau e^{-0.00061\sigma_0^2} \tag{3.111}$$

3.8.4 Frequency Scaling of Rain-Induced Cross-Polarization

Adopting the frequency scaling model of rain-induced cross-polarization, we can predict XPD_2 value of another frequency and polarization angle by using long-term statistical result XPD_1 of a certain frequency and polarization angle. XPD_1 and XPD_2 correspond to the same time probability. The scaling formula is [27]:

$$XPD_2 = XPD_1 - 20\lg\left[\frac{f_2\sqrt{1-0.484(1+\cos 4\tau_2)}}{f_1\sqrt{1-0.484(1+\cos 4\tau_1)}}\right] \tag{3.112}$$

In the formula: f_1, τ_1 and f_2, τ_2 refer to different frequencies and polarization angles.

Formula (3.112) is applicable to $4 \le f_1, f_2 \le 30$. Since the applicable frequency of formula (3.100) is between 6 and 55 GHz, *XPD* at 4–6 GHz can be obtained by 6 GHz prediction result in combination with formula (3.112).

3.9 Other Propagation Effects

3.9.1 Wet Antenna Attenuation

During a rainfall event, water is accumulated on the reflector of the antenna, antenna housing, as well as horn hood causing attenuation of satellite signals. At the same time, the receiving quality of the antenna also reduces owing to the defocusing caused by the refraction of radio waves produced by the water film on the antenna and surface of the feed resource. The additional attenuation caused by the rainfall mentioned above is collectively called as wet antenna attenuation. According to several experiments, wet antenna attenuation increases with an increase in rain attenuation and finally reaches a maximum value; antenna water-accumulation attenuation of 20 GHz and 27 GHz earth-space links is up to 6 dB and 8 dB, respectively. Since there is limited test data of wet antenna attenuation, and wet antenna attenuation is relative to elevation angle of the antenna, antenna type, rainfall type, wind speed, and direction, it is challenging to develop a model of wet antenna attenuation. In fact, wet antenna attenuation is included in the forecasting model of rain attenuation of earth-space links.

3.9.2 Attenuation Due to Sand and Dust

The mechanism of attenuation caused due to sand and dust is identical to that of attenuation due to clouds and rain. In the microwave frequency band, the attenuation due to dust particles can be calculated using Rayleigh scattering theory, while in millimeter wave band and above, Mie's theory is more fitting. With limitations imposed by test data, studies on attenuation due to sandstorm are still in the theoretical stage, signifying that the attenuation rate of sandstorms is inversely proportional to the visibility of the sandstorm and is closely related to the water content in sand. In case of dry sandstorms with visibility of 100 m, the attenuation rates at 14 GHz and 37 GHz are 0.03 dB/km and 0.15 dB/km, respectively. For sandstorms with water content of 20% and the same particle size distribution, the attenuation rates are about 0.65 dB/km and 1.5 dB/km, respectively. Based on existing studies, the effect of sandstorm on a satellite's signals is small for the Ka frequency band and below, while the sandstorm with high concentration and high water content can produce considerable attenuation of a satellite's signals for 30 GHz frequency band and above.

3.10 Total Atmospheric Attenuation

For satellite communication systems operating at the Ka frequency band and above, it is necessary to consider the overall influence produced by atmospheric gas attenuation, scintillation fading, cloud attenuation, and in particular, rain attenuation. The prediction method of total atmospheric attenuation with p (%) time probability exceeded provided by ITU-R is as follows [27]:

$$A_T(p) = A_G(p) + \sqrt{(A_R(p) + A_C(p))^2 + A_S^2(p)} \tag{3.113}$$

When $P\% < 1\%$,

$$A_C(p) = A_C(1) \tag{3.114}$$

$$A_G(p) = A_G(1) \tag{3.115}$$

In the formula: $A_R(p)$, $A_C(p)$, $A_G(p)$, and $A_S(p)$ represent rain attenuation, cloud attenuation, gas attenuation, and scintillation attenuation, respectively, with $p(\%)$ time probability exceeded.

3.11 Radio Wave Environment Data in Prediction Model

The radio wave environment behaves as an assembly of space environment factors that affect radio wave propagation. The radio wave environment of the troposphere is also called as radio meteorology environment. The prediction reliability of radio wave propagation depends on the credibility of the surrounding environment parameters imported into the model, to a huge extent. Traditional radio meteorology parameters typically adopt partition parameters and empirical models, which are mostly inaccurate and could lead to discontinuity of prediction result of partition boundary. To improve precision and reliability of radio wave environment parameters, universally, people make use of long-term ground and sounding meteorology observation data and evaluate meteorology data to create a digital map of the radio wave environment, including atmospheric temperature, humidity, air pressure, atmospheric refractive index, refractive index gradient, rainfall rate, cloud integration water content, and integration vapor content, to enhance the accuracy of radio wave environment parameters.

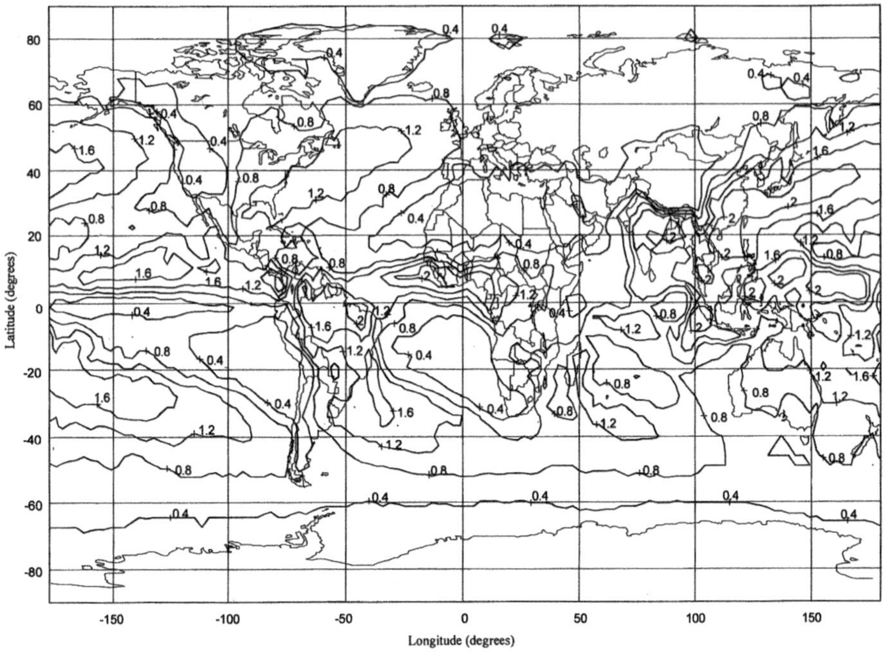

Fig. 3.25 Example of ITU-R's digital maps [53]. (Normalized total columnar content of cloud liquid water exceeded for 1% of the year)

3.11.1 ITU-R Digital Maps of Radio Environment

ITU-R established various digital maps of radio wave environment parameters essential for propagation forecasting based on reanalysis data ERA-15 (1978–1994) in the European Center for Medium-Range Weather Forecasts, wherein, the radio wave environment parameters related to the forecasting of earth-space path propagation characteristics are as shown in Fig. 3.14. In these digital maps [33], the space position is exported by several rows and columns, and the space relationship is implied by illustrating an adjacent relation in the mesh. Therefore, locating the parameters at a certain longitude/latitude becomes incredibly convenient when we know the starting point, size, and the number of rows and columns of the mesh. Figure 3.25 provides the global statistical map of rainfall rate with 0.01% time probability exceeded and vapor integration content with 1% time probability exceeded (Table 3.14).

3.11.2 China's Digital Map of Radio Environment

China Research Institute of Radio Propagation utilized the long-term observation data from 1991 to 2000 of more than 1,000 surface meteorological stations and more

Table 3.14 ITU-R's digital map of main radio wave environment related to forecasting of earth-space path propagation characteristics [33]

ITU-R's recommendation	Data variable	Mesh accuracy	Spatial interpolation methods Time probability interpolation method		Variable interpolation method
P.839	Height of zero layer	1.5° × 1.5°	Bilinear interpolation	–	–
P.837	Cumulative distribution of rainfall rate	0.25° × 0.25°	Bilinear interpolation	Semi-logarithmic interpolation	–
P.1511	Elevation	0.5° × 0.5°	Bicubic interpolation	–	–
P.836	Vapor column integration content	1.125° × 1.125°	Bilinear interpolation	Semi-logarithmic interpolation	Linear interpolation
P.836	Vapor density of earth's surface	1.125° × 1.125°	Bilinear interpolation	Semi-logarithmic interpolation	Linear interpolation
P.1510	Annual average temperature of earth's surface	0.75° × 0.75°	Bilinear interpolation	–	–
P.453	Refractive index wet item	0.75° × 0.75°	Bilinear interpolation	–	–
P.840	Cloud integration water content	1.125° × 1.125°	Bilinear interpolation	Semi-logarithmic interpolation	Linear interpolation

than 100 weather sounding stations located in China and the surrounding regions, to set up 11 types of Chinese digital maps of the radio wave environment parameters of the troposphere together with the distribution features of China's observation stations.

Compared to ITU-R's digital maps, Chinese digital maps can reflect China's climatic features in a more effective manner. Figure 3.26 provides an example of such digital maps.

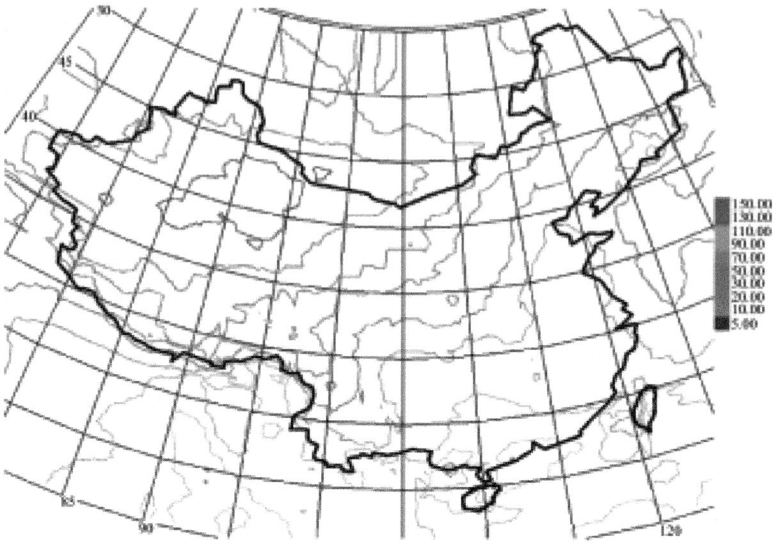

(a)

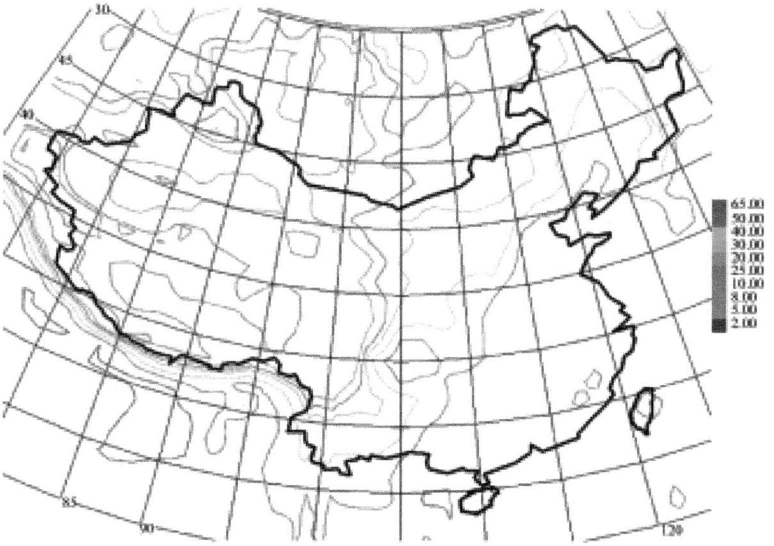

(b)

Fig. 3.26 Example of Chinese digital maps. **a** Rainfall with 0.01% time probability exceeded; **b** integration vapor content with 1% time probability exceeded

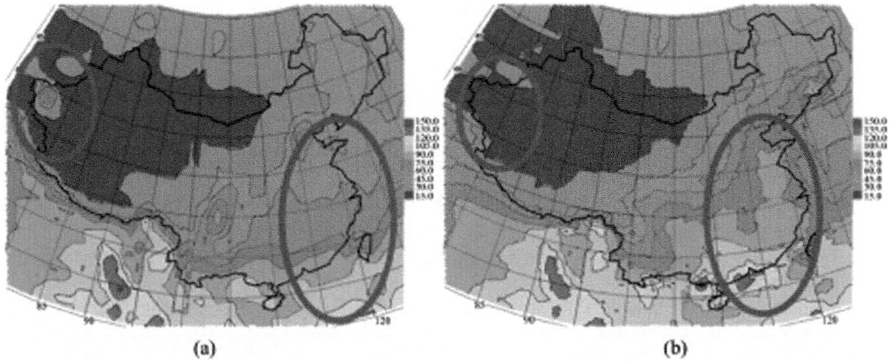

Fig. 3.27 Comparison of digital maps of rainfall rate with time probability exceeded. **a** Generated by ITU-R; **b** generated in accordance with China's data

3.11.3 Selection Principle of Radio Environment Parameters

The accuracy of radio environment parameters directly impacts the result of propagation forecasting. Radio wave environment parameters obtained via different data resources naturally tend to be different, and thus, reasonable selection of the parameters can boost the accuracy and reliability of propagation forecasting. When performing radio wave propagation effect forecasting in China, the parameters can be obtained using the following sequences:

(1) Obtain cumulative distribution required by environment parameters in accordance with the local long-term climate monitoring data (more than 10 years).
(2) Obtain environment parameters of the required regions through interpolation in accordance with the digital maps of radio wave environment generated by China's long-term climate monitoring data (more than 10 years).
(3) Obtain environment parameters of the required regions through interpolation by using ITU-R's digital maps of radio wave environment.

Figure 3.27 depicts a comparison of the ITU-R's digital map of rainfall rate and the Chinese digital map of rainfall rate with 0.01% time probability exceeded, which is generated using China's statistical data. ITU-R's rainfall rate distribution basically decreases from south to north with latitude, and there exists a high rainfall rate distribution center in the Kashi Region and the Xinjiang Uygur Autonomous Region, which contrasts with the real status quo.

References

1. Jinshang Z, Changsheng L, Wenbao L (1962) Basis of atmospheric physics. China Meteorological Press, Beijing
2. Prupacher HR, Pitter RL (1971) A semi-empirical determination of the shape of clouds and rain drops. J Atmos Sci 28:86–94
3. Gunn R, Kinzer GD (1949) The terminal velocity of fall for water droplets in stagnant air. J Meteorol 6:243–248
4. Laws JO, Parsons DA (1943) The relation of raindrop size to intensity. Trans Am Geophys Union 24:452–460
5. Marshall JS, Palmer WM (1948) The distribution of raindrops with size. J Meter 5:165–166
6. Joss J, Thams JC, Waldvogel A (1968) The variation of raindrop size distributions at Lacarno. In: Proceedings of the international coference on cloud physics, Toronto, pp 369–373
7. Shengbo Q, Jinghua C (1995) Raindrop size distribution in Guangzhou. Chin J Radio Sci 10(4):73–77 (in Chinese)
8. Zhenwei Z (1995) Raindrop size distribution model and rain attenuation forecast in Guangzhou. Chin J Radio Sci 10(4):33–37
9. Riva C (2002) Spatial characteristics of propagation parameters: a review. COST Action 280
10. Feral L (2003) HYCELL-a new hybrid model of the rain horizontal distribution for propagation studies: 1. Modeling of the rain cell. Radio Sci 38(3)
11. Feral L (2003) HYCELL-A new hybird model of the rain horizontal distribution for propagation studies: 2. Statistical modeling of the rain rate field. Radio Sci 38(3)
12. REC. ITU-R P.837-6 (2012) Characteristics of precipitation for propagation modelling
13. Shengbo Q, Jinghua C (1997) The conversion formulate of rainfall rate distributions with different integration times at the typical regions in China. Chin J Radio Sci 12(1):112–117
14. Leke L, Zhenwei Z, Yumei L (2002) Unified conversion model of rainfall rates at different integral times in China. Chin J Radio Sci 17(6):641–645
15. ITU-R Document 3J/59-E (1999) Conversion of rain rate statistics to various rain gauge integration times. ITU-R Study Group meeting, Geneva
16. Ray PS (1972) Broadband complex refractive indices of ice and water. Appl Opt 11(8):1836–1844
17. Debye P (1929) Polar molecules. Dover, New-York
18. Manabe TH, Liebe J, Hufford GA (1987) Complex permitivity of water between 0 and 30 THz. In: 12th international conference on infrared and millimeter waves, Lake Buena Vista, FL, December 1987, pp 14–18
19. Ulaby FT, Moore RK, Fung AK (1981) Microwave remote sensing, vol 1. Addison-Westley Publishing Company
20. Oguchi T (1981) Scattering from hydrometeors: a survey. Radio Sci 16(5):691–730
21. Oguchi T (1983) Eletromagnetic wave propagation and scattering in rain and other hydrometerors. Proc IEEE 71(9):1029–1078
22. REC. ITU-R P.840-8 (2019) Attenuation due to clouds and fog
23. Zhenwei Z, Zhensen W, Guangde S et al (2000) An empirical model for calculating cloud millimeter wave attenuation. Chin J Radio Sci 15(3):300–303
24. Zhenwei Z, Wu Z (2000) Millimeter-wave attenuation due to fog and clouds. Int J Infrared Millim Waves 21(10):1607–1616
25. REC. ITU-R P.838-3 (2005) Specific attenuation model for rain for use in prediction methods
26. REC. P.841-5 (2016) Conversion of annual statistics to worth-month statistics [S]. Geneva. ITU-R
27. REC. ITU-R P.618-13 (2017) Propagation data and prediction methods required for the design of earth-space telecommunication systems
28. Lu C-S, Zhao Z-W, Wu Z-S et al (2018) A new rain attenuation prediction model for the earth-space links. IEEE Trans Antennas Propag
29. REC. ITU-R P.581-3 (2022) The concept of "worst month" [S]. Geneva. ITU-R

30. REC. ITU-R P.1623-1 (2005) Prediction method of fade dynamics on earth-space paths
31. REC. ITU-R P.1853-1 (2012) Tropospheric attenuation time series synthesis
32. Zhenwei Z (2012) The two-order small argument approximation of rain-induced XPD theory and XPD prediction. Chin J Radio Sci 27(5):52–58
33. REC. ITU-R P.1144-9 (2017) Guide to the application of the propagation methods of radiocommunication study group 3

Chapter 4
Earth's Surface Reflection and Local Environment Effects

The rapid development of advanced technologies like satellite communication, navigation, remote sensing, and Non-Geostationary Earth Orbit (Non-GEO) satellite systems with extensive application in onshore, aerial, and maritime mobile terminals make it essential for mobile satellite service systems to comprehensively consider their propagation characteristics, to maintain reliability. Due to the constant variation in elevation angle and application environment of radio wave propagation path of a mobile satellite service, the mobile terminal generally uses the wide-beam antenna, which is capable of receiving the incoming-wave signals from different directions, including direct signals, reflected signals of ground and sea surface, as well as reflected and scattering signals of various objects. Furthermore, satellite signals could also be shadowed by the buildings and trees along the roadsides. For this reason, the mobile satellite system, unlike the fixed satellite system, must consider the environment effects, including ground reflection and local environment features, in addition to the propagation effect of fixed satellite services.

4.1 Overview

The mobile satellite system essentially delivers mobile satellite services, such as communication and navigation, to terminal users. Theoretically, only 3 GEO satellite systems are capable of acquiring the global coverage, minus the places near polar regions, through the proper configuration. The Inmarsat system, managed by International Maritime Satellite Organization (Inmarsat, renamed International Mobile Satellite Organization) [1], aims to provide communication services for global maritime, land, and aerial users via the GEO satellite system. The position of 3rd generation satellite—Inmarsat-3 is shown in Fig. 4.1. Inmarsat-3 is capable of dynamically allocating the power and frequency owing to the introduction of a point beam and dual-polarization technique, thereby tremendously boosting the

Z. Zhao et al., *Radio Wave Propagation of Satellite Systems*,
https://doi.org/10.1007/978-981-97-1394-3_4

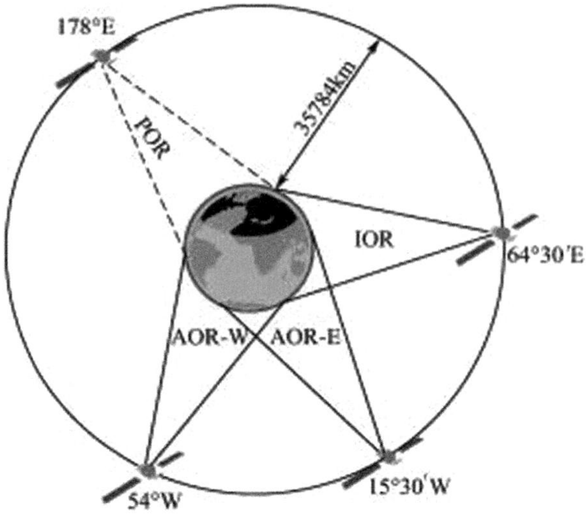

Fig. 4.1 Inmarsat-3 satellite position. (Quoted from http://commons.wikimedia.org/wiki/Image:Satellite_coverage.jpg)

utilization rate of satellite channel resources. Inmarsat group keeps launching new satellites regularly; 3 Inmarsat-5 satellites operating at the Ka frequency band were launched in 2013 and 2015.

Large signal attenuation and long propagation delays form the chief disadvantages of the application of GEO satellites in the mobile satellite business. To evade these issues, majority of the mobile satellite businesses make use of Low Earth Orbit (LEO) or Medium Earth Orbit (MEO) satellites, such as Iridium, Global star, GPS, and BDS, among which, the Iridium satellite system [2], consisting of 66 LEO communication satellites, delivers mobile communication business worldwide. This system has 6 orbital planes and 11 satellites are positioned on each plane with an orbital inclination of 86°, which are polar orbit satellites. The Iridium system is as shown in Fig. 4.2.

Depending on the location where the mobile ground stations are located, the global mobile satellite service is divided into maritime mobile satellite service, land mobile satellite service, and aeronautical mobile satellite service. The environment where the user terminal is located and how it gets affected by the propagation effects differently with different mobile satellite businesses. The maritime mobile satellite service is predominantly affected by sea refection, but land mobile satellite service experiences heavy impact of local environments, such as surrounding mountains, plants, and buildings, in addition to ground surface reflection; especially, the signals of direct wave also undergo shadowing by plants and buildings suffering severe attenuation.

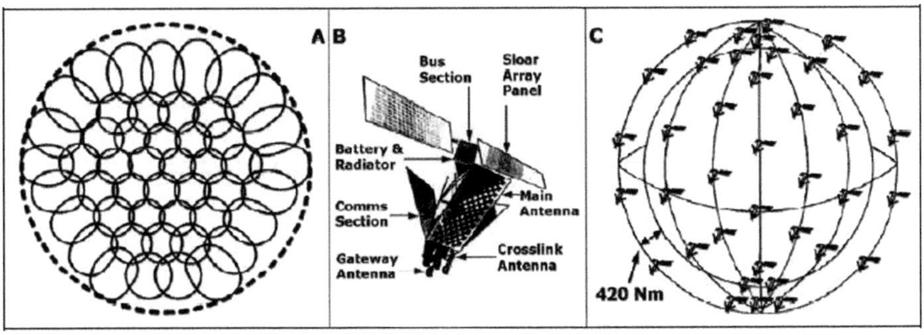

Fig. 4.2 Iridium spot coverage, iridium satellite, and space constellation [2]

Table 4.1 Fade depth measured in an experiment

Time probability/%	0.1	1	5	10	30	50	90	99	99.9
Fade depth/dB	31	25.46	18.97	8.15	2.312	0.1072	−0.3646	−1.209	−2.327

Due to earth's surface reflection, signals arrive at the receiver from different directions and along multiple paths causing overlap of the direct wave signal, mirror reflection signal, and diffuse reflection signal, resulting in multipath fading of the receiving signal. Moreover, multipath propagation also gives rise to intersymbol interference, which deteriorates the communication quality drastically. Multipath propagation, shadowing, and blockage effects of the local environment on the direct-path signal lead to severe and rapid fading of the receiving signal at the ground station, which is the principal aspect affecting the performance of a satellite system. The intensity of signal fading is always characterized by fade depth. Fade depth exceeded for $p\%$ time or distance is defined by the ratio between the direct wave signal level and the signal level exceeded for $(1-p\%)$ time, usually expressed in decibels.

Table 4.1 Provides the measured fade depth in an experiment.

In this chapter, the fading predicting method of multipath and shadowing effects in maritime, land, and aeronautical mobile satellite services, along with the influence of rain attenuation on mobile satellite services are mainly discussed.

4.2 Radio Wave Propagation Effects of Maritime Mobile Satellite Service

The maritime mobile satellite service provides communication and navigation services for large- and middle-sized ocean vessels, coastal and inland-water small-sized ships, fishing vessels, yachts, and lifeboats, in addition to naval vessels and offshore drilling platforms. For a century, the maritime mobile satellite system has delivered great assistance for maritime trade and rescue. This mobile satellite service experiences severe performance degradation due to sea surface reflection and scattering, and such an effect proves more detrimental for wide-beam antennas. The fading effect due to sea surface reflection and the predicting method of interference effect between the adjacent satellites are described in this section [3].

4.2.1 Fading Due to Sea Surface Reflection

The user terminal of maritime mobile satellite service is located on the sea, and its receiving signals are comprised of direct wave signal and sea surface reflection signal, among which, the reflection signal in turn also contains a mirror reflection component (coherent component) and diffuse reflection component (incoherent component), while the synthesized signal amplitude is consistent with Nakagami-Rice distribution (also referred to as Rice distribution) [4]. The probability density function of Rice distribution is

$$p(x) = \frac{x}{\sigma^2} e^{-\frac{x^2+a^2}{2\sigma^2}} \mathrm{I}_0\left(\frac{xa}{\sigma^2}\right) \tag{4.1}$$

where $2\sigma^2$ is the average power of the random vector (W), a^2 is the power of fixed vector (W), $I_0(.)$ is a modified Bessel function of the first kind and of zero order. If the sum of fixed vector power and random vector power is unity 1, that is, $a^2 + 2\sigma^2 = 1$, assuming α is the proportion of random vector in total power, that is, $\alpha = 2\sigma^2$, then the Rice distribution curves with α as parameter and 1 as total power are shown in Fig. 4.3.

The subsequent paragraphs introduce the predicting method of sea surface reflection fading [3], which can be applied to the condition of 0.8 GHz $\leq f \leq$ 8 GHz of frequency, $5° \leq \theta \leq 20°$ of elevation angle, circular polarization, and 1–3 m of height of sea waves. Assuming the direct wave signal power is 0 dB, then the average incoherent power of sea surface reflection wave P_r (dB), corresponding to the direct wave, can be represented as

$$P_r = G + R + \eta_I \quad \text{dB} \tag{4.2}$$

where G is the relative antenna gain in mirror reflection direction (dBi), calculated using Formula (4.3); R is the amplitude of reflection coefficient for circularly polarized wave (dB), calculated using Formula (4.4); η_I is the standard diffuse reflection coefficient (dB), obtained from Fig. 4.4.

$$G = -4 \times 10^{-4}\left(10^{G_m/10} - 1\right)(2\theta)^2 \tag{4.3}$$

$$R = 20\lg|R_C| \tag{4.4}$$

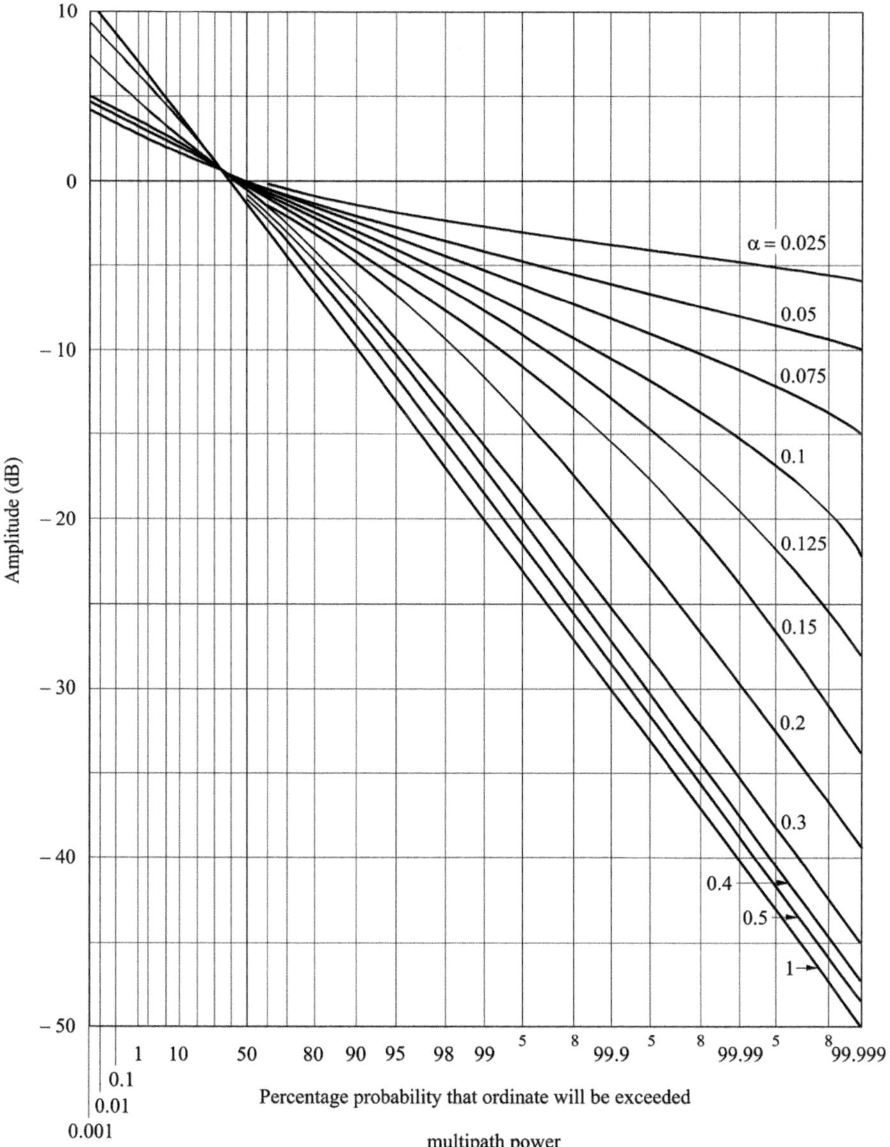

$$\alpha = \frac{\text{multipath power}}{\text{total power}}$$

$$\text{For this case, } \alpha = \frac{10^{P_r/10}}{1 + 10^{P_r/10}}$$

Fig. 4.3 Rice distribution of constant total power (parameter is α) [5]

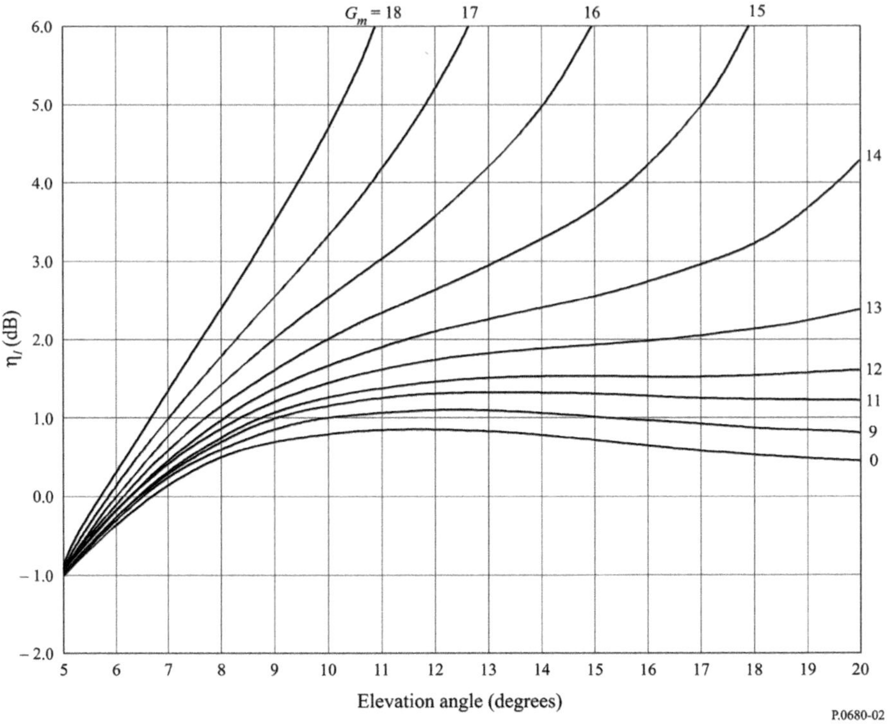

Fig. 4.4 Average normalized diffuse reflection coefficients in the frequency range 0.8–8 GHz [3]

where G_m is the maximum gain of antenna (dB), R_c is the Fresnel reflection coefficient of the sea for circular polarization, which can be defined as

$$R_C = \frac{R_H + R_V}{2} \tag{4.5}$$

where R_H and R_V are the reflection coefficient of horizontally polarized wave and vertically polarized wave, respectively, which can be defined as

$$R_H = \frac{\sin\theta - \sqrt{\eta - \cos^2\theta}}{\sin\theta + \sqrt{\eta - \cos^2\theta}} \tag{4.6}$$

$$R_V = \frac{\sin\theta - \sqrt{(\eta - \cos^2\theta)/\eta^2}}{\sin\theta + \sqrt{(\eta - \cos^2\theta)/\eta^2}} \tag{4.7}$$

where η is the complex relative permittivity, $\eta = \varepsilon_r(f) - j60\lambda\sigma(f)$; ε_r is the relative permittivity of sea surface; σ is the conductivity of sea surface (S/m) [6]; f is the frequency (GHz); λ is the free-space wave length (m).

For the condition that power of direct wave signal is 0 dB and average power of sea surface reflection wave is P_r, then the parameter α of Rice distribution can be represented as

$$\alpha = \frac{10^{P_r/10}}{1 + 10^{P_r/10}} \tag{4.8}$$

Fade depth exceeded for $p\%$ time is calculated using the following formula:

$$F_d(p) = -\left[A + 10\lg\left(1 + 10^{P_r/10}\right)\right] \tag{4.9}$$

where A is the amplitude (dB) read from the ordinate of Fig. 4.3, and the time probability of A exceeded is $(1-p\%)$ in accordance with the definition of the fade depth.

4.2.2 Prediction of Interference from Adjacent Satellites

In the maritime mobile satellite communication system, the desired signal from the satellite and the interfering signal from an adjacent satellite undergo independent signal level fluctuation. In general, interference from an adjacent satellite is predicted by the calculation of parameters, such as signal-to-noise ratio, signal-to-interference ratio, and signal-to-interference plus noise ratio. Overall, two types of interferences from adjacent satellite systems are observed: One is the downlink interference on the mobile earth station terminal, and the other is the uplink interference on the satellite terminal, as shown in Fig. 4.5. Another situation is interference between beams in multi-spot-beam operation, where the same frequency is allocated repeatedly.

The prediction of interference [3, 4] from an adjacent satellite is based on three assumptions: ① The amplitude of desired signal as well as the interference signal has the Rice distribution, or the interference signal has the Rayleigh distribution,

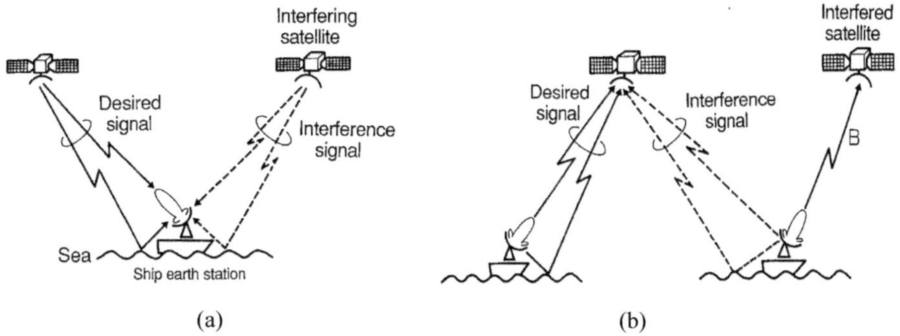

Fig. 4.5 Interference from adjacent satellites [6]. **a** Downlink interference, **b** uplink interference

while the amplitude variation of these two signals is purely independent. ② Thermal noise does not vary with time. ③ When interference signal and thermal noise occur at the same time, the equivalent noise power is equal to the sum of interference power and thermal noise power. Based on the above assumptions, the interference from an adjacent satellite can be predicted using the following method:

The input parameters can be defined as follows in units of power, not dB.

D—power of direct wave component of desired signal
M—average power of the reflection component of desired signal
N—average power of system noise
I_D—direct wave component power of interference signal
I_M—average power of reflection component of interference signal
I—average power of interference wave, $I = I_D + I_M$.

All the above parameters adopt linear values of power instead of dB values.

Signal-to-noise ratio [c/n] (p) is the ratio of power of the desired signal to system noise power as an function of time percentage p and is expressed by

$$[c/n](p) = (\eta_c)^2(p)D/N \tag{4.10}$$

where η_c is the normalized time percentage dependent factor of desired signal power having a Rice distribution. Assuming direct wave signal power is 0 dB, then η_c can be expressed as

$$20\lg \eta_c = A + 10\lg((D + M)/D) \tag{4.11}$$

where A is the amplitude (dB) read from the ordinate of Fig. 4.3, and in the figure parameter $\alpha = M/(D + M)$ for this case. Signal-to-interference ratio [c/i] (p) is the ratio of power of the desired signal to power of the interference signal, being expressed as

$$[c/i](p) = \left(\eta_{c/i}\right)^2(p)D/I_{50} \tag{4.12}$$

where I_{50} is the median value of power variation of the interference signal, that is, the value of 50% time probability.

$$I_{50} = \left(\eta_{i,50}\right)^2 I \tag{4.13}$$

$\eta_{c/i}(p)$ is the normalized time percentage dependent factor of signal-to-interference ratio [c/i]:

$$\left[\lg \eta_{c/i}(p)\right]^2 = \left[\lg \eta_c(p)\right]^2 + \left[\lg \eta_i(100 - p)\right]^2 \tag{4.14}$$

where η_i is the normalized time percentage dependent factor of interference signal power. The probability density function of $\eta_i \eta_{i,50}$ has a Rice distribution, in which,

Table 4.2 Relationship between $\eta_{i,50}$, η_i, and b

b	$1_M/1_D$/dB	$\eta_{i,50}$ /dB	η_i /dB							
			50%	20%	10%	5%	1%	0.5%	0.1%	0.01%
0	∞	−1.59	0.00	3.66	5.21	6.36	8.22	8.83	9.98	11.25
0.5	0	−1.12	0.00	3.16	4.48	5.44	7.03	7.54	8.52	9.60
0.6	−1.8	−0.91	0.00	2.88	4.09	4.99	6.46	6.95	7.87	8.90
0.7	−3.7	−0.68	0.00	2.53	3.62	4.43	5.78	6.22	7.08	8.03
0.8	−6.0	−0.45	0.00	2.10	3.03	3.72	4.90	5.30	6.07	6.92
0.9	−9.5	−0.22	0.00	1.52	2.21	2.76	3.69	4.00	4.62	5.32
0.95	−12.8	−0.11	0.00	1.09	1.61	2.02	2.74	2.99	3.48	4.02
1.0	−∞	−0.00	0.00	0.00	0.00	0.00	0.00	0.00	0.00	0.00

the total signal power is 0 dB, 20lg($\eta_i \eta_{i,50}$) is the amplitude (dB) read from Fig. 4.3, and parameter $\alpha = I_M/I$. $\eta_{i,50}$ is the correction factor, 20lg $\eta_{i,50}$ is the value observed in Fig. 4.3 corresponding to 50% time probability; due to the effect of correction factor $\eta_{i,50}$, $\eta_i = 1$ when $p = 50$. If $\eta_c < 1$ and $\eta_i > 1$, $\eta_{c/i} < 1$ should be selected, while $\eta_{c/i} > 1$ must be selected in other situations. By setting $I_D/I = b$, then the relationship between $\eta_{i,50}$, η_i, and b is given in Table 4.2.

$[c/(I + n)](p)$ is the ratio of power of desired signal to power of noise plus interference signal, being expressed as

$$[c/(i+n)](p) = \left[1/[c/n](p) + 1/[c/i](p)\right]^{-1} \tag{4.15}$$

When N ≤ −5 dB, M ≤ −5 dB, I ≤ −10 dB, and 0.5 ≤ b ≤ 1, the prediction accuracy of the Formula (4.15) lies within 1 dB, in which, N, M, and I are normalized with respect to direct wave component D of desired signal, that is, assume D = 1 W or D = 0 dB.

4.3 Radio Wave Propagation Effects of Land Mobile Satellite Service

Land mobile satellite service offers communication, navigation, and other telecommunication services to numerous ground users, especially in regions that cannot be covered by terrestrial mobile communication systems, such as remote country sides, mountainous areas, and so on. The user terminal of land mobile satellite service always employs omni-directional receiving antennas, and it is imperative to consider the severe effects of multipath, shadowing, and blockage produced by the terrain, buildings, trees, etc., on system performance. Availability of land mobile satellite service is usually less than the fixed satellite system, because generally the users of the service are located in more convoluted and changing environment. In general, the optimized availability range of system design is between 80 and 90%. The propagation effect of multipath, shadowing, and blockage of land mobile satellite service and their prediction methods are introduced in this section.

4.3.1 Multipath Model for Line-of-Sight Conditions

In the majority of cases, the mobile terminal has a clear line-of-sight condition with respect to the satellite. At this point, signal fading caused by the ground multipath still takes place. A phasor summation of the line-of-sight signal received by the mobile terminal with the multipath signal resulting in an increase or decrease of the amplitude of the receiving signal. The multipath fading is closely related to parameters such as polarization mode and receiving antenna pattern, in addition to cross section and quantity, and distance from the receiving antenna of the multipath reflectors.

Next, we introduce the multipath propagation prediction model under the mountainous and roadside vegetation environment. The model involved is based on the measured data; the characteristics of the measured antennas are: The azimuth of antennas is omni-directional; gain change of 15°–75° of elevation angle is smaller than 3 dB; antenna gain below the horizon decreases by 10 dB as a minimum.

1. Multipath fading in mountainous environment:

In a mountainous landscape, fade depth distribution caused by multipath can be expressed as [7, 8]:

$$p = aA^{-b} \tag{4.16}$$

In the formula: p is the percentage of distance over which the fading is exceeded ($1\% < p\% < 10\%$); A refers to fading value exceeded (dB); a and b denote fitting parameters, and their values are listed in Table 4.3.

Figure 4.6 provides the cumulative distribution curve of fading at 1.5 GHz and 870 MHz with 30° and 45° of elevation angle of path.

2. Multipath fading in a roadside vegetation environment:

According to various tests and research conducted in the United States, for tree-lined roads, multipath fading is not sensitive to elevation angle when the elevation angle

Table 4.3 Optimized fitting parameters of the cumulative fading distribution of multipath environment in a mountainous terrain

Frequency/GHz	Elevation angle 30°			Elevation angle 45°		
	A	B	Range/dB	a	b	Range/dB
0.87	34.52	1.855	2–7	31.64	2.464	2–4
1.5	33.19	1.710	2–8	39.95	2.321	2–5

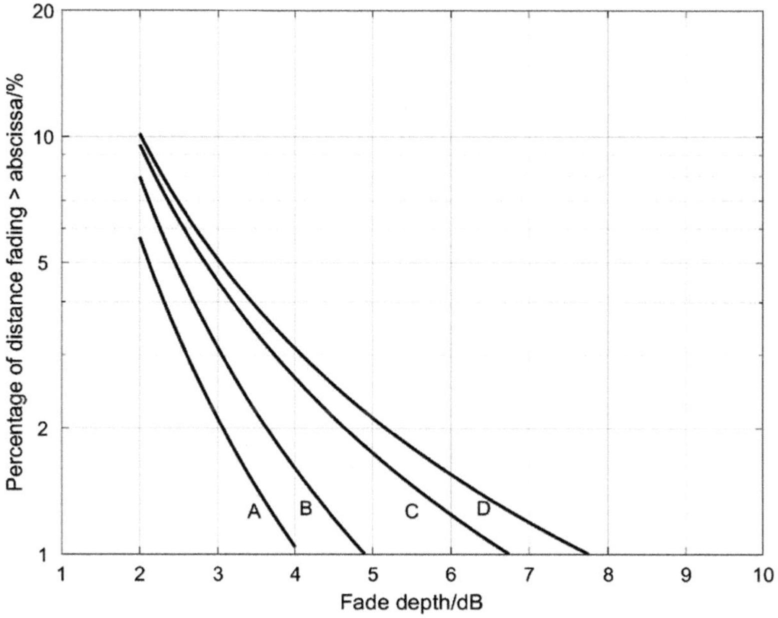

Fig. 4.6 Cumulative fading distribution of multipath fading in a mountainous terrain. Curve A-870 MHz, 45° Curve B-1.5 GHz, 45° Curve C-870 MHz, 30° Curve D-1.5 GHz, 30°

of the path is 30° to 60°. Assuming that the shadowing effect of the trees on a direct wave can be overlooked, the multipath fading caused by the trees can be expressed as [7]:

$$p = u\exp(-vA) \tag{4.17}$$

In the formula: p refers to the percentage of distance over which the fading is exceeded ($1\% < p\% < 50\%$); A refers to the fading value exceeded (dB); fitting parameters u and v are listed in Table 4.4.

Figure 4.7 provides 870 MHz and 1.5 GHz cumulative fading distribution curve.

Enhanced fading due to multipath can occur at lower elevation angles (5°–30°) where forward scattering from relatively smooth rolling terrain can be received from larger distances.

Table 4.4 Optimized fitting parameters of the cumulative multipath fading distribution of tree-lined roads

Frequency/GHz	u	v	Fading range/dB
0.870	125.6	1.116	1–4.5
1.5	127.7	0.8573	1–6

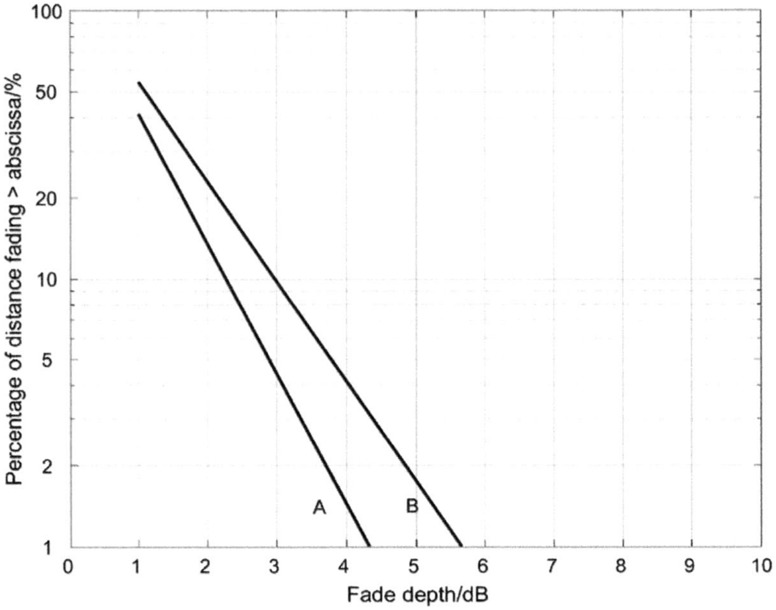

Fig. 4.7 Cumulative fading distribution of multipath fading of tree-lined roads. Curve A-870 MHz; Curve B-1.5 GHz

4.3.2 *Shadowing Effects*

1. Shadowing effect of trees:

The empirical prediction model for shadowing effect of trees along the roadside is obtained based on the measurement results of 870 MHz, 1.6 GHz, and 20 GHz fading cumulative distribution. While measuring, the vehicle runs in a lane in both the directions to measure shadowing effect of trees at different elevation angles. The prediction method of such fading distribution is applicable for highways and rural roads, with the assumption that the shadowing effect of trees is the primary cause of fading of satellite signals. This shadowing fading model [7] can be applied to the condition of 800 MHz–20 GHz frequency, 7°–60° elevation angle, and 1%–80% distance percentage.

(1) Firstly, use the following formula to calculate the fading distribution at 1.5 GHz of frequency, $1\% \leq p\% \leq 20\%$ distance percentage, and $20° \leq \theta \leq 60°$ elevation angle:

$$A_L(p, \theta) = -M(\theta)\ln(p) + N(\theta) \tag{4.18}$$

where

$$M(\theta) = 3.44 + 0.0975\theta - 0.002\theta^2 \tag{4.19}$$

$$N(\theta) = -0.443\theta + 34.76 \tag{4.20}$$

(2) In case of 0.8 GHz $\leq f \leq$ 20 GHz frequency, 1% $\leq p\% \leq$ 20% distance percentage, and 20° $\leq \theta \leq$ 60° elevation angle, the fading distribution can be converted using the following formula:

$$A_{20}(p, \theta, f) = A_L(p, \theta) \exp\left\{1.5\left[\frac{1}{\sqrt{f_{1.5}}} - \frac{1}{\sqrt{f}}\right]\right\} \tag{4.21}$$

(3) In case of 0.8 GHz $\leq f \leq$ 20 GHz frequency, 20% $\leq p\% \leq$ 80% distance percentage, and 20° $\leq \theta \leq$ 60° elevation angle, the formula of fading distribution is as follows:

$$A(p, \theta, f) = A_{20}(20\%, \theta, f)\frac{1}{\ln 4}\ln\left(\frac{80}{p}\right) \tag{4.22}$$

Figure 4.8 illustrates the fading values at 1.5 GHz for elevation angle (10° $\leq \theta \leq$ 60°) and distance percentage (1% $\leq p\% \leq$ 50%).

When frequency is 1.6 and 2.6 GHz, the roadside tree shadowing model can be extended to elevation angle θ > 60° using linear interpolation with the following procedure. Formulae (4.18) to (4.22) can be used to calculate the fading value of

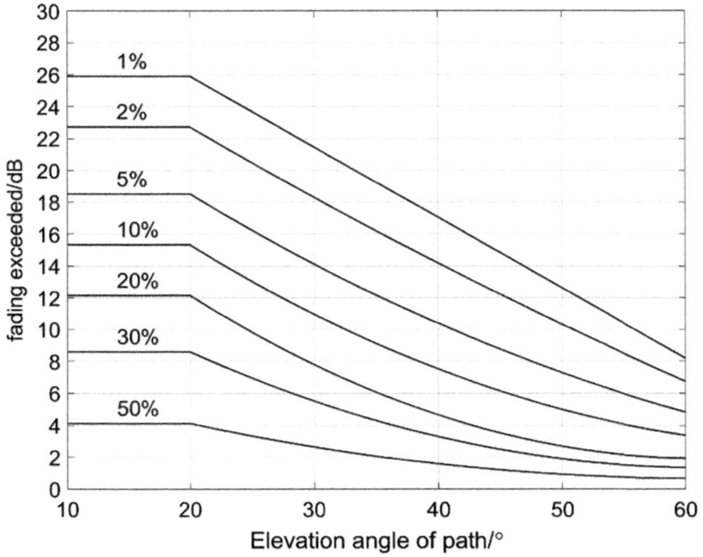

Fig. 4.8 Relationship between roadside shadow fading and elevation angle at 1.5 GHz

Table 4.5 Fading value of 80° elevation angle

p/(%)	Vegetation shadowing fading	
	1.6 GHz	2.6 GHz
1	4.1	9.0
5	2.0	5.2
10	1.5	3.8
15	1.4	3.2
20	1.3	2.8
30	1.2	2.5

60° elevation angle, while obtaining the fading value of 80° elevation angle from Table 4.5, and setting the fading value of 90° elevation angle to be 0; if elevation angle is $60° < \theta < 80°$, linearly interpolate based on 60° and 80° elevation angle; if elevation angle is $80° < \theta < 90°$, linearly interpolate based on 80° and 90° elevation angle.

The prediction methods mentioned above are based on the assumption of a fixed elevation angle. For mobile satellite systems, the elevation angle of the path can be varied, and the following procedures must be considered to calculate link reliability.

① Calculate the time probability of each elevation angle or angle scope when the mobile terminal is located in a line-of-sight range of the satellite;
② For a given propagation margin (ordinate in Fig. 4.8), get unavailability probability at each elevation angle; if the antenna of the mobile terminal is not omni-directional, it is essential to deduct antenna gain from propagation allowance;
③ For each elevation angle, multiply the result of procedure ① with the result of procedure ② and then divide by 100 to determine unavailability probabilities under this angle;
④ Add all the unavailability probabilities obtained in procedure ③ to get the unavailability of the whole system.

2. Shadowing effect of roadside buildings:

When the mobile user terminal is located on urban roads, the satellite signal gets shadowed and blocked by the surrounding buildings. For easy modeling of the shadowing effect of buildings, the height distribution of buildings can be modeled by a Rayleigh distribution. Figure 4.9 is the geometric sketch of the shadowing effect of buildings.

The blocking probability due to the buildings is [7]:

$$p = 100\exp\left[-(h_1 - h_2)^2/2h_b^2\right] \quad h_1 > h_2 \tag{4.23}$$

In the formula: h_1 is height of ray above the ground in front of the building; h_2 is Fresnel clearance required above buildings; h_b is building height.

Calculation formula of h_1 and h_2 is shown as below:

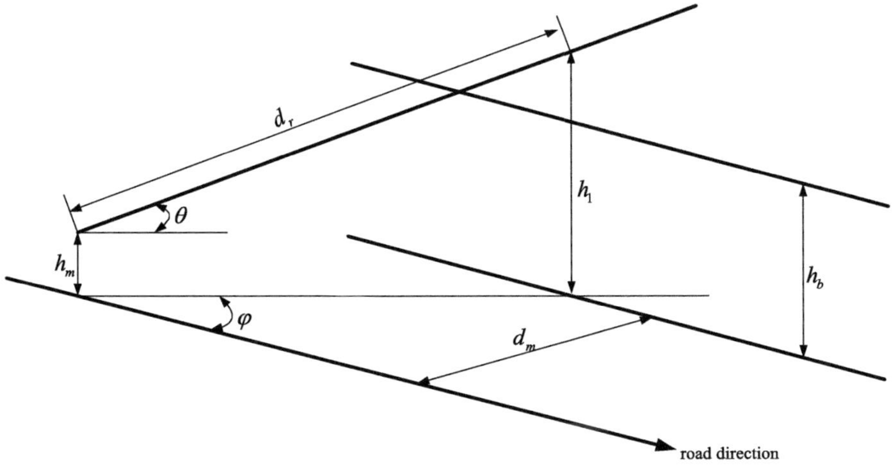

Fig. 4.9 Geometric sketch of the shadowing effect of buildings

$$h_1 = h_m + (d_m \tan\theta / \sin\varphi) \tag{4.24}$$

$$h_2 = C_f (\lambda d_r)^{0.5} \tag{4.25}$$

In the formula: h_m is the height above ground of mobile station; d_m is distance of mobile station from the front of the building; θ is elevation angle (°) of satellite to horizontal line; φ is azimuth angle of ray to street direction (°); C_f is the portion of required clearance in first Fresnel zone; λ is wave length; d_r is slope distance from mobile station to a position along the ray vertically above the front of the building, being expressed as

$$d_r = d_m / (\sin\varphi \cdot \cos\theta) \tag{4.26}$$

In Formulae (4.23) to (4.26), the parameters such as h_1, h_2, h_b, h_m, d_m, d_r, and λ are in self-consistent units, and $h_1 > h_2$. Note that the Formulae (4.24), (4.25), and (4.26) are valid under the condition of $0 < \theta < 90°$ and $0 < \varphi < 180°$. Figure 4.10 demonstrates the shadowing effect of roadside buildings calculated using the above formula with $h_b = 15$ m, $h_m = 1.5$ m, $d_m = 17.5$ m, and 1.6 GHz of frequency.

Although this model indicates that no blocking occurs at high path elevation, the user may still experience occasional shadowing and blockage when passing through an overpass, a suspended road sign, branches, etc. Additionally, for portable mobile terminals, propagation would be affected due to the shadowing caused by the head and the body of the humans.

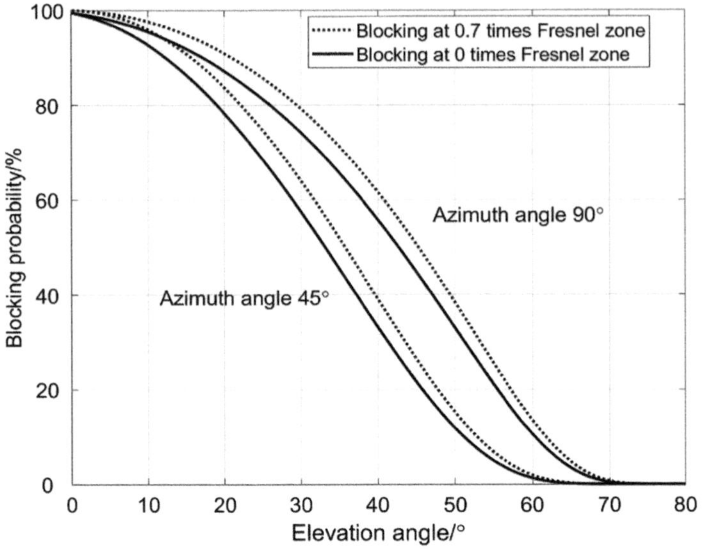

Fig. 4.10 Example of shadowing of roadside buildings

4.4 Radio Wave Propagation Effect of Aeronautical Mobile Satellite Service

The aeronautical mobile satellite service primarily offers communication and navigational services for military and civil aircraft. To ensure air safety, a highly reliable propagation channel is required. A single flying event of an aircraft includes stages, such as take-off, cruising, and landing, and the position where the aircraft is located at each stage is significantly different, therefore, the aeronautical mobile satellite service must consider the different propagation effects depending on the respective characteristics. Furthermore, multipath fading due to earth's surface reflection is unlike the fading due to other mobile satellite services, since an aircraft's speed and height are much greater than any other mobile user platform. As small-sized antennas are typically used on an aircraft, The performance of antennas is affected by the dynamic change of the airframe to further induce the propagation effect of signals. With consideration of air safety, the short-term signal fading cannot be overlooked. Radio wave propagation effect of aeronautical mobile satellite service under different environments and its prediction method are introduced in this section.

4.4.1 Fading Due to Sea Surface Reflection

The fading characteristic prediction of aeronautical mobile satellite service is similar to fading due to sea surface reflection of maritime satellites, and here, it is crucial to consider the influence of earth's curvature since the aircraft flies at a high altitude. The influence of earth's curvature increases with an increase in flying altitude.

With fading due to sea surface reflection of aeronautical mobile service, the level amplitude of synthesized signal also has a Rice distribution, and its probability density function refers to Formula (4.1). The prediction method of fading [9] due to sea surface reflection in this section is applicable under the following conditions: 1 GHz $\leq f \leq$ 2 GHz of frequency; for circularly polarized wave and horizontally polarized wave, elevation angle of satellite is $\theta \geq 3°$, and antenna gain is $G(1.5\theta) \geq -10$ dB, for vertical polarization, $\theta \geq 8°$; wave height is 1–3 m.

Assuming direct wave signal power is 0 dB, then, the average incoherent power of sea surface reflection wave P_r, corresponding to the direct wave, can be represented as

$$P_r = G + R + C_\theta + D \quad \text{dB} \tag{4.27}$$

where G is relative antenna gain (dB) in mirror reflection direction; R is amplitude (dB) of different polarization reflection coefficient; C_θ is the correction factor (dB); and D is diffusion factor due to earth's curvature (dB).

Parameters G, R, C_θ, and D in Formula (4.27) can be calculated using the following formula:

$$\begin{cases} G = -4 \times 10^{-4}\left(10^{G_m/10} - 1\right)(\theta_i)^2 \\ \theta_i = \theta + \left(\theta_{sp} + \theta_{hr}\right)/2 \\ \theta_{sp} = 2\gamma_{sp} + \theta \\ \theta_{hr} = \cos^{-1}[R_e/(R_e + H_a)] \end{cases} \tag{4.28}$$

$$\begin{cases} R = 20\lg|R_i| \\ R_i = R_H, R_V, R_C \end{cases} \tag{4.29}$$

$$C_\theta = \begin{cases} 0 & \theta_{sp} \geq 7° \\ \left(\theta_{sp} - 7\right)/2 & \theta_{sp} < 7° \end{cases} \tag{4.30}$$

$$D = -10\lg\left[1 + \frac{2\sin\gamma_{sp}}{\cos\theta_{sp}\sin\left(\gamma_{sp} + \theta_i\right)}\right] \tag{4.31}$$

where G_m is max gain value (dB) of antenna gain; θ_i is the angle (°) deviating from visual axis; θ_{sp} and θ_{hr} are grazing angle (°) and line-of-sight angle (°) of mirror reflection point, respectively; $\gamma_{sp} = 7.2 \times 10^{-3}\ H_a/\tan\theta$; R_e is the earth's radius, R_e = 6,371 km; H_a is the antenna height (km).

Fade depth exceeded for p% time is calculated using the following formula:

$$F_d(p) = -\left[A + 10\lg\left(1 + 10^{P_r/10}\right)\right] \tag{4.32}$$

where A is the amplitude (dB) read from the ordinate of Fig. 4.3, and the time probability A exceeded is $(1-p\%)$ in accordance with the definition of fade depth.

4.4.2 Multipath Model for Aircraft During Approach and Landing

The aeronautical communication and navigation system should focus on the influence of multipath propagation with short delay, in particular, at the stages of the aircraft's approach and landing, since the availability and reliability of the communication, and accuracy and integrity of the navigation services are highly significant. It is known that the reflection of ground and fuselage can produce considerable propagation effects as well.

The multipath signal received by the aircraft mainly consists of reflected signal from the structure of the aircraft itself and reflected signal of the ground. When the antenna is arranged at the top of cockpit, it only produces strong reflection at the fuselage, rather than the wings; so basically, this reflected signal with a short delay does not vary with time and dominates the channel. Ground reflection shows very high time-varying characteristic, and its Doppler frequency shift is relative to the descent speed of an aircraft.

The multipath model [9] of the aircraft approach and landing can be used for frequencies between 1 and 3 GHz. The satellite azimuth can vary between 10° and 170°, or 190° and 350°. The elevation angle to the satellite can vary between 10° and 75°. Figure 4.11 demonstrates the model of the complete aeronautical channel during the aircraft's approach, which mainly consists of four paths: The first path is line-of-sight component; the second path is flat fading part of line-of-sight component; the third path is multipath fading process due to fuselage with 1.5 ns of delay; the fourth path is terrain echo, whose delay is decided by elevation angle and altitude. The input parameters of the model include azimuth angle $\varphi(t)$ of satellite, elevation angle $\theta(t)$ of satellite, and altitude $h(t)$ of the aircraft above the ground. In these, t represents the time, with each parameter being a function of time.

The model requires the knowledge of the aircraft geometry and flight dynamics. The model parameters vary with different aircraft types. Next, we consider the jet aircraft VFW614 (ATTAS, representing small-sized jet aircraft) and airbus A340 (representing large-sized commercial aircraft) as examples to introduce the aeronautical channel model, as shown in Table 4.6.

1. Direct path

In addition to the line-of-sight component (path 1), this path is rigorously modulated, and its amplitude satisfies the Rice distribution. The generation of the fading process refers to the Formulae (4.33, 4.34, 4.35).

2. Reflection of fuselage

To get the time series of the reflected signal of the fuselage, its power spectrum density must be known. Time series can be derived by filtering complex Gaussian white noise, where the corresponding function of the filter is the power spectral density function that can be calculated using the following formula:

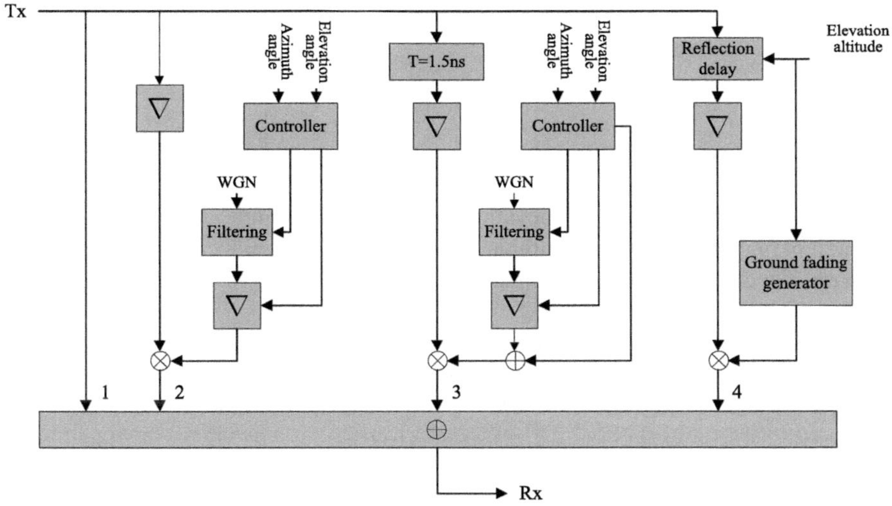

Fig. 4.11 Model of complete aeronautical channel

Table 4.6 Parameters of channel model

Parameter	Delay/ns	Relative power/dB	Doppler bandwidth/Hz
Direct path			
Mean component	0	0	0
Fading process		(−14.2-mean value)	<0.1
Reflection path of fuselage			
Mean component	1.5	−14.2	
Fading process		(−14.2-mean value)	<0.1
Reflection path of ground			
Ground reflection	900–10 (Descent process)	−15 to −25	<20 (Deviation due to descent speed)

$$p_{\text{proc}}(\text{dB}) = b_1 + b_2 \cdot e^{b_3 \cdot |f|} \tag{4.33}$$

where b_2 and b_3 are the coefficients of the exponential process.

In addition to the noise process, the reflected signal of the fuselage also contains a −14.2 dB mean (DC) component, and constant b_1 (dB) is defined by the following formula:

$$b_1 = -14.2 - mean \quad \text{dB} \tag{4.34}$$

Coefficient *mean* satisfies the following fourth-order two-dimensional polynomial:

$$mean(\theta,\varphi) = \begin{bmatrix}\theta^4 & \theta^3 & \theta^2 & \theta & 1\end{bmatrix} \cdot A_{mean} \cdot \begin{bmatrix}\varphi^4 \\ \varphi^3 \\ \varphi^2 \\ \varphi \\ 1\end{bmatrix} \tag{4.35}$$

where *mean* is the function of elevation angle θ and azimuth angle φ. A_{mean} is the coefficient matrix of a 5 * 5 polynomial. Coefficient b_2 and b_3's calculation is similar to *mean*. For ATTAS, coefficient matrix of *mean*, b_2, and b_3 is shown as follows:

$$A_{mean,ATTAS} = \begin{bmatrix} -2.0057e{-12} & 5.0499e{-10} & -4.6114e{-8} & 1.8053e{-6} & -2.4773e{-5} \\ 2.8598e{-10} & -7.4259e{-8} & 7.0553e{-6} & -2.9116e{-4} & 0.0043 \\ -1.1568e{-8} & 3.2474e{-6} & -3.3846e{-4} & 0.0156 & -0.2698 \\ 3.8681e{-8} & -2.2536e{-5} & 0.0038 & -0.2512 & 6.3140 \\ 1.9434e{-6} & -3.5747e{-4} & 0.0133 & 0.8133 & -28.1329 \end{bmatrix}$$

$$A_{b2,ATTAS} = \begin{bmatrix} -3.9148e{-11} & 8.8672e{-9} & -7.0048e{-7} & 2.2069e{-5} & -2.1492e{-4} \\ 6.0699e{-9} & -1.3708e{-6} & 1.0784e{-4} & -0.0034 & 0.0322 \\ -3.2203e{-7} & 7.2344e{-5} & -0.0057 & 0.1747 & -1.6206 \\ 6.7649e{-6} & -0.0015 & 0.1162 & -3.5328 & 31.6814 \\ -4.4741e{-5} & 0.0098 & -0.7383 & 21.9981 & -142.3524 \end{bmatrix}$$

$$A_{b3,ATTAS} = \begin{bmatrix} -1.8398e{-12} & 4.2182e{-10} & -3.3813e{-8} & 1.0855e{-6} & -1.0875e{-5} \\ 2.6665e{-10} & -6.0897e{-8} & 4.8490e{-6} & -1.5346e{-4} & 0.0015 \\ -1.2870e{-8} & 2.9171e{-6} & -2.2947e{-4} & 0.0071 & -0.0629 \\ 2.3542e{-7} & -5.2520e{-5} & 0.0040 & -0.1193 & 0.9153 \\ -1.2058e{-6} & 2.5797e{-4} & -0.0187 & 0.5027 & -4.1128 \end{bmatrix}$$

For A340, coefficient matrix of *mean*, b_2, and b_3 is shown as follows:

$$A_{mean,A340} = \begin{bmatrix} -2.6220e{-12} & 6.0886e{-10} & -5.0686e{-8} & 1.8074e{-6} & -2.3633e{-5} \\ 4.3848e{-10} & -1.0231e{-7} & 8.6113e{-6} & -3.1465e{-4} & 0.0044 \\ -2.3577e{-8} & 5.5538e{-6} & -4.7815e{-4} & 0.0184 & -0.2872 \\ 3.9552e{-7} & -9.2657e{-5} & 0.0082 & -0.3431 & 6.9937 \\ -1.5225e{-6} & 3.3690e{-4} & -0.0312 & 1.7110 & -32.8066 \end{bmatrix}$$

$$A_{b2,A340} = \begin{bmatrix} -3.1880e{-11} & 7.2724e{-9} & -5.8454e{-7} & 1.9069e{-5} & -1.9707e{-4} \\ 4.7229e{-9} & -1.0775e{-6} & 8.6761e{-5} & -0.0028 & 0.0293 \\ -2.3471e{-7} & 5.3437e{-5} & -0.0043 & 0.1413 & -1.4541 \\ 4.4756e{-6} & -0.0010 & 0.0812 & -2.6731 & 27.5448 \\ -2.5361e{-5} & 0.0056 & -0.4459 & 14.8917 & -109.1083 \end{bmatrix}$$

$$A_{b3,A340} = \begin{bmatrix} -1.2021e{-12} & 2.7780e{-10} & -2.2626e{-8} & 7.4413e{-7} & -7.5120e{-6} \\ 1.7647e{-10} & -4.0725e{-8} & 3.3131e{-6} & -1.0855e{-4} & 0.0011 \\ -8.6470e{-9} & 1.9871e{-6} & -1.6099e{-4} & 0.0052 & -0.0488 \\ 1.6123e{-7} & -3.6656e{-5} & 0.0029 & -0.0946 & 0.8204 \\ -8.5647e{-7} & 1.8942e{-4} & -0.0149 & 0.4826 & -5.5011 \end{bmatrix}$$

Table 4.7 Altitude regions of the Markov model

Altitude classification	Start altitude /m	End altitude /m
High	1000	400
Middle	400	100
Low	100	10

3. Ground reflection

The Doppler frequency shift of ground reflection is relative to the vertical descent speed $V_{\text{vert}}(t)$ of the aircraft, and its mean Doppler frequency shift is

$$f_g(t) = \frac{v_{vert}(t)}{\lambda} \tag{4.36}$$

where λ is the wavelength (m).

Under the mean Doppler frequency shift calculated using the Formula (4.36), the Doppler spectrum of ground reflection can be expressed by the normalized Gaussian distribution:

$$P_{Gr} = P_g + 20\lg\left(\frac{1}{\sigma\sqrt{2\pi}} \cdot e^{-\frac{f^2}{2\sigma^2}}\right) \tag{4.37}$$

where P_g is the ground reflection power (dB) obtained by Markov model; standard deviation $\sigma = 2.92$ (Hz), that can be obtained through tests.

To establish a model of ground reflection [10], the aircraft approach process is divided into three altitude regions, such as "high," "middle," and "low," based on the flying height. The ground reflection in each region is described using the Markov model. The definitions of altitude regions are provided in Table 4.7 and Fig. 4.12.

The receiving power of aircraft is divided into four levels or states, and the output power of each state is tabulated in Table 4.8 (state 1 does not consider ground reflection).

The Markov transition probability is obtained by using the measurement data. The matrix P of conversion probability is defined independently at each altitude region, where $P_{x,\,y}$ denotes the probability of converting state x to state y. The transition probability obtained through measurement data of the landing process in the Australian Graz airport is shown as follows:

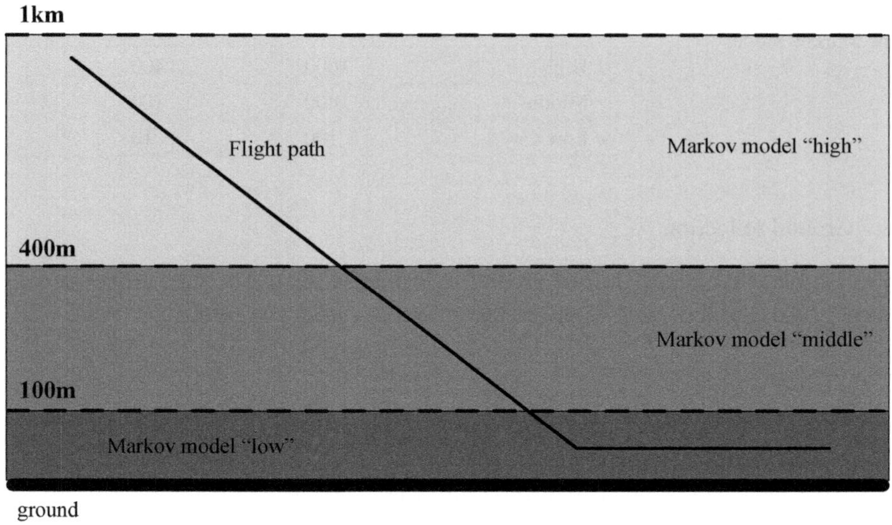

Fig. 4.12 Altitude regions of ground reflection model

Table 4.8 Ground fading states of the Markov model

State	Power /dB
1	<−25
2	−23
3	−19
4	−15

$$
\begin{aligned}
P_{400-1000} &= \begin{bmatrix} 0.9866 & 0.0087 & 0.0047 & 0 \\ 0.6087 & 0.3043 & 0.0870 & 0 \\ 0.2143 & 0.3571 & 0.4286 & 0 \\ 0.3333 & 0.3333 & 0.3334 & 0 \end{bmatrix} \\
P_{100-400} &= \begin{bmatrix} 0.9842 & 0.0130 & 0.0028 & 0 \\ 0.6667 & 0.2222 & 0.0889 & 0.0222 \\ 0.0667 & 0.1167 & 0.5000 & 0.3166 \\ 0 & 0 & 0.3279 & 0.6721 \end{bmatrix} \\
P_{10-100} &= \begin{bmatrix} 0.9645 & 0.0310 & 0.0045 & 0 \\ 0.7308 & 0.1538 & 0.1154 & 0 \\ 0.6250 & 0.1250 & 0.2500 & 0 \\ 0.3333 & 0.3333 & 0.3334 & 0 \end{bmatrix} \\
P_{0-10} &= \begin{bmatrix} 1 & 0 & 0 & 0 \\ 1 & 0 & 0 & 0 \\ 1 & 0 & 0 & 0 \\ 1 & 0 & 0 & 0 \end{bmatrix}
\end{aligned}
\tag{4.38}
$$

In the formula: P_{x-y} refers to transition probability at altitude region $h(t) \geq x$ and $h(t) < y$. The frequency of state transition is 25.4 Hz, that is, output power is required to change 25.4 times in 1 s.

Consequently, the process of ground reflection fading is as shown in Fig. 4.13. Firstly, define the Markov transition probability matrix P based on flying altitude to get the output power value, where the transition speed is 25.4 Hz, and the output power is denoted as P_g in Formula (4.37); secondly, calculate Doppler spectrum using Formula (4.37), the fading time series can be obtained from complex white Gaussian noise by Doppler filtering; thirdly, get mean Doppler frequency shift using Formula (4.36) in accordance with the aircraft's descent speed and wavelength; finally, multiply these parameters to obtain the fading process of the ground path.

Assume that the terrain near the airport is flat, then it is very easy to calculate the delay time $T_g(t)$ of the ground reflection path with a function of the elevation angle of the path:

$$\tau_g(t) = \frac{2 \cdot h(t) \cdot \sin(\theta)}{c} \tag{4.39}$$

In the formula: $T_g(t)$ refers to delay time (s) of ground reflection path; c refers to velocity of light (m/s); $h(t)$ refers to aircraft altitude (m); θ refers to elevation angle (°).

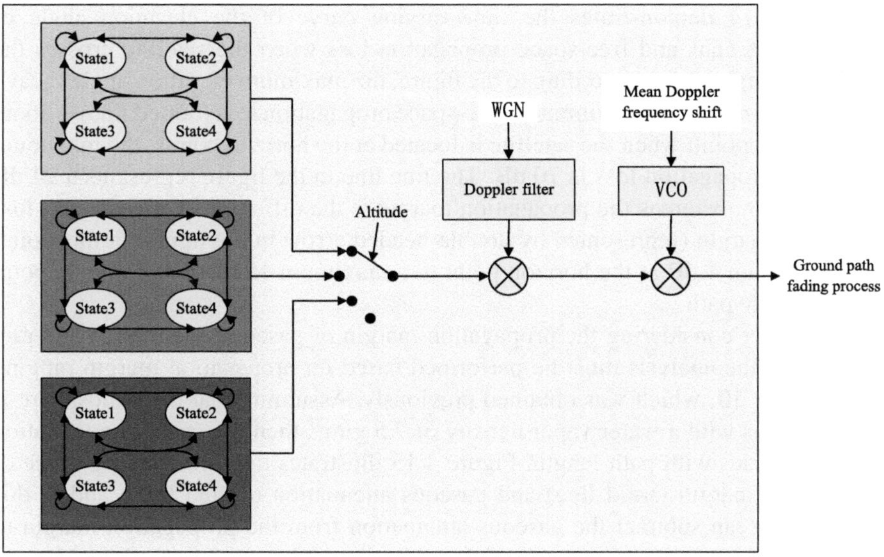

Fig. 4.13 Fading process of ground path

4.5 Analysis of Propagation Effect for Non-GEO Satellite Service

Non-GEO satellite communication service has undergone a rapid stride of progress in the past few decades. For a non-GEO, its path length is variable when it passes through the ground terminal. The path tends to be longest when the satellite rises and falls at the horizon line, and is shortest when the satellite passes through the middle point. Therefore, the statistical characteristics of the elevation angle of the path must be considered during the propagation prediction process to acquire the proper propagation margin. Since non-GEO communication service has been extended to the Ka frequency band and above, it is vital to consider the influence of rain attenuation on non-GEO links. The complexity of the evaluation of a non-GEO satellite propagation effect is that the slant path of a satellite is a time-varying parameter instead of a fixed parameter. Now, we consider an example to illustrate the evaluation process of non-GEO link propagation effect [11].

Consider one low earth orbit satellite (LEO) in a circular polar orbit at an altitude of 765 km, and the rising node is at 100° west longitude. Assuming that the satellite operates at the *Ka* frequency band, with the user terminal located at 106.6° west longitude and 32.5° northern latitude, its downlink works at 20 GHz, and its power margin is set at 74dB available for free-space propagation loss along and other propagation losses.

Figure 4.14 demonstrates the time-varying curve of the elevation angle of receiving antennas and free-space propagation loss when the satellite crosses the terminal a single time. According to the figure, the maximum elevation angle (heavy line) is about 60°, and the minimum free-space propagation loss (dotted line) is about 59 dB at this point; when the satellite is located at the horizon points, the maximum free-space propagation loss is 70 dB. The fine line in the figure represents a 74 dB power margin, whereas the propagation margin is the difference between path loss and power margin (represented by double-headed arrow in the figure) at this point, from minimum 4 dB at the horizon point to a maximum 15 dB at the middle point of the satellite path.

For further considering the propagation margin of gaseous attenuation and rain attenuation, the analysis must be performed based on propagation margin ranging from 4 to 15 dB, which was obtained previously. Assuming that the atmosphere is homogeneous with a water vapor density of 7.5 g/m^3, then the gaseous attenuation [12] only varies with path length. Figure 4.15 illustrates the time-varying curve of propagation margin (solid line) and gaseous attenuation (dotted line), and at this moment, we can subtract the gaseous attenuation from the propagation margin to obtain the rain attenuation margin of the system (represented by the dotted line in the figure). Rain attenuation margin varies with time and reaches up to 14.7 dB (max value) at the middle point of the satellite path, but changes to a negative value in the first and the last 2 min of the process. This is indicative of the occurrence of system interruption in case of rain at low elevation angles.

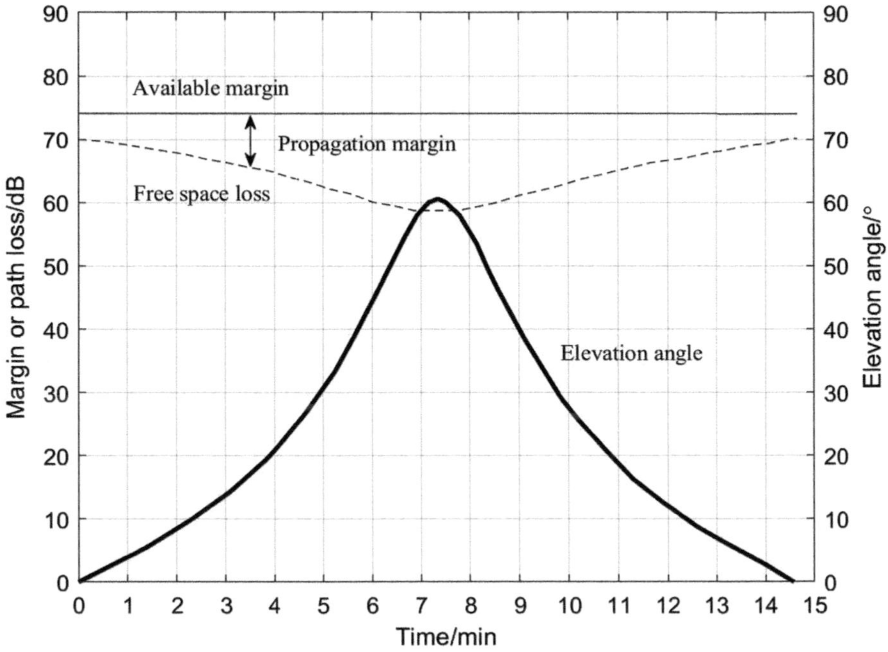

Fig. 4.14 Elevation angle and path loss during single-crossing of LEO satellite (20 GHz of frequency and at an altitude of 765 km)

Based on the above, the relative reliability of various parts during the communication of the ground station with the LEO satellite can be evaluated in accordance with rainfall statistics and rain attenuation forecasting mode of the ground station.

Figure 4.16 provides the annual interruption probability during the communication between ground station and LEO satellite. Based on the result, the annual probability of outage of the link (99.99% link availability) is less than 0.01% at minutes 6, 7, and 8 at the mid-point when the satellite is passing, but the link availability is reduced to less than 99% in the initial and final 2 min. The satellite communication could be interrupted at 1 min near the horizon points.

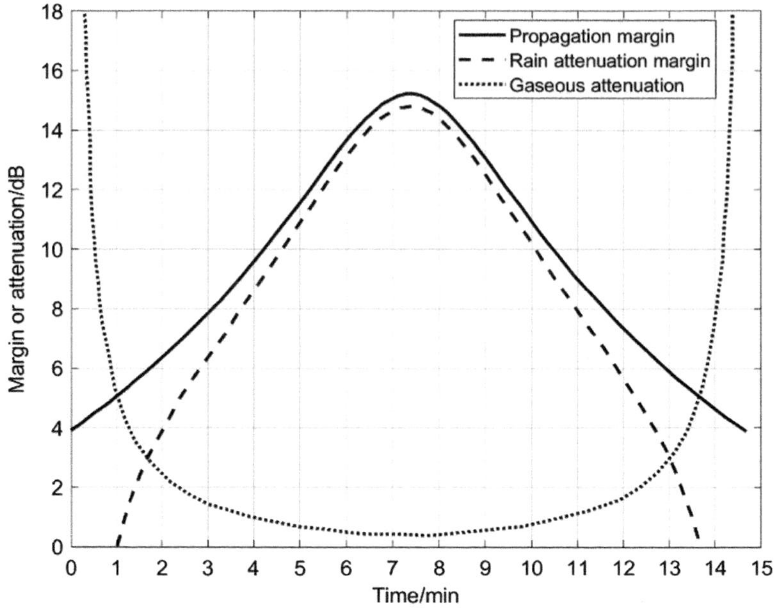

Fig. 4.15 Propagation margin, gaseous attenuation, and rain attenuation margin (20 GHz) for a single passing of LEO satellite

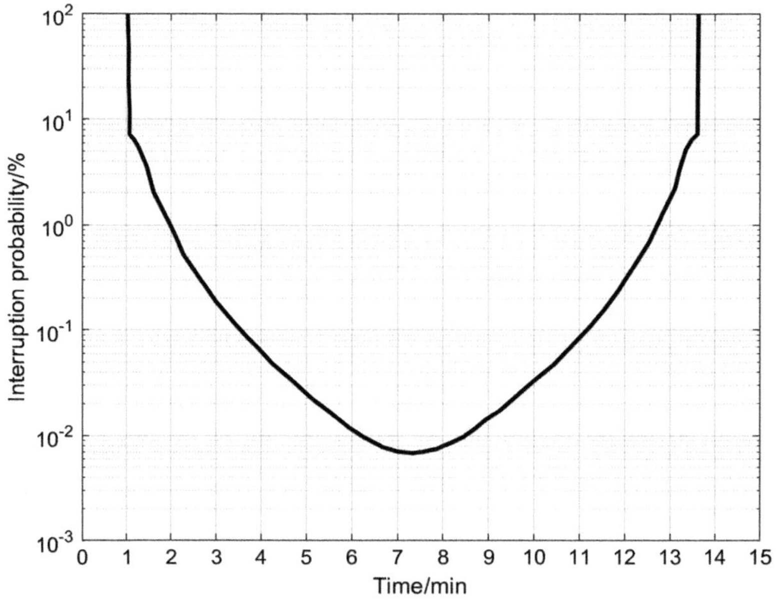

Fig. 4.16 Interruption probability of LEO satellite links (20 GHz)

References

1. SD Ilčev (2005) Global mobile satellite communications: for maritime, land and aeronautical applications. Springer Science & Business Media
2. "Technical background", Iridium, Washington (1999)
3. Rec. ITU-R P.680-3 (1999) Propagation data required for the design of Earth-space maritime mobile telecommunication systems
4. Karasawa Y, Yasunaga M (1992) Interference evaluation method for mobile-satellite systems under Nakagami-Rice fading conditions. IEICE Trans Commun 75:42–49
5. ITU-R WP3M (1996) Handbook on radiowave propagation information for predictions Earth-space communications.
6. Rec. ITU-R P.527-3 (1992) Electrical characteristics of the surface of the earth
7. Rec. ITU-R P.681-10 (2017) Propagation data required for the design of Earth-space land mobile telecommunication systems
8. Goldhirsh J, Vogel WJ (1998) handbook of propagation effects for vehicular and personal mobile systems. NASA reference publication
9. Fiebig UC, Artaoud G, Issler JL, et al (2009) Channel modelling activities related to the satellite navigation channel in the SatNEx project. In: 3rd European Conference
10. Rec. ITU-R P.682-3(2012), Propagation data required for the design of Earth-space aeronautical mobile telecommunication systems
11. ITU-R Doc.3M/111 (2005) Propagation data required for the design of Earth-space aeronautical mobile telecommunication systems
12. Ippolito LJ (1999) propagation effects handbook for satellite systems design

Chapter 5
Ionospheric Effects

The ionosphere layer behaves as a diffusedly ionized medium exhibiting significant temporal and spatial variability, which is essentially characterized on the scales of diurnal, seasonal, and 11-year solar periodic variation. Ionosphere is known to generate a series of propagation effects on radio waves, such as absorption, refraction, delay, dispersion, polarization rotation, and scintillation, which are strong contributory factors toward the performance of satellite systems operating at 3 GHz and below. In this chapter, the characteristics of ionospheric environment and its chief propagation effects are introduced.

5.1 Ionospheric Characteristics

5.1.1 Overview of the Ionosphere

UV rays and X-rays of the solar radiation ionize the high-level atmosphere partially forming a weak plasma medium, i.e., the ionosphere at the height of 60–1,000 km. Extensive temporal and spatial variation of the ionosphere results in propagation effects of the satellite signals, such as absorption, refraction, delay, and polarization rotation. The small-scale ionospheric irregularities induce random fluctuations in the phase/density of the receiving signal and the angle of arrival, giving rise to scintillation fading of the satellite signals. The performance of navigation, remote sensing, and communication system is critically influenced by the said propagation effects.

The free electron concentration in a unit volume of ionosphere, namely, electron density, acts as the primary determining factor for radio wave propagation. The variation in composition of the gas and density at different altitudes causes the ionized solar radiation lines or frequency bands to be distinct as well, giving rise to the apparent layering feature of the ionosphere. The ionospheric electron density is

Z. Zhao et al., *Radio Wave Propagation of Satellite Systems*,
https://doi.org/10.1007/978-981-97-1394-3_5

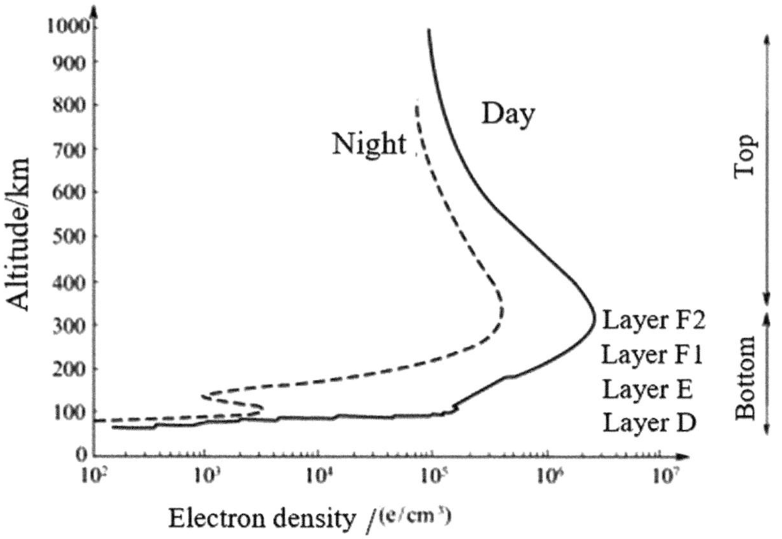

Fig. 5.1 Structural sketch of the layering of ionosphere [1]

distributed vertically, starting from 60 km above ground, and exhibits several peak values with the increase in altitude, and subsequently portrays a reduction with the constant increase in altitude. Depending on the heights of the peak value regions of electron density in the ionosphere, it is usually divided into layer D, layer E, layer F1, and layer F2, as shown in Fig. 5.1. This figure illustrates the height and diurnal variation of the ionospheric layers. The ionosphere is significantly characterized by diurnal variation under the influence of solar radiation. The electron density in layer D and layer E is relatively low, while that in layer F2 is detectably large. At night, the electron density is evidently lower than that during the day, and while the ionization level is minimal in layer D, it is mostly absent in layer F1 at night, making it fade away.

Main characteristics of the ionospheric layers are shown as follows:

Layer D: 60–90 km, the density is about 10^8–10^{10} e/m^3 during the day and disappears at night. The neutral atmosphere composition density in layer D is quite large. The electrons frequently collide with neutral particles to form anions, thereby the ion density in layer D is higher than electron density.

Layer E: 90–150 km, electron density is about 10^9–10^{11} e/m^3. Layer E is relatively stable.

The height of Layer E electron density peak is always around 110–120 km in mid-latitude regions but decreases by 10 km in low-latitude regions.

Layer F1: 150–200 km, electron density is about 10^{11} e/m^3 and fades away or merges with layer F2 at night. Its features are similar to layer E, and the diffusion movement is not very apparent at this height.

Table 5.1 Physical features of ionospheric layers

Region name	Region D	Region E	Region F1	Region F2
Region scope	60–90	90–150	150–200	200–500
Height at max electron density/km	≈70	≈110	180–200	≈300
Max electron density Nmax/(e.m^3)	10^8–10^{10}	10^9–10^{11}	10^{11}	10^{11}–10^{12}
Basic features	Disappears at night	Electron density is higher during the day and lower at night	Electron density in layer F_1 fades away at night, which usually occurs in summer season; electron density in layer F_2 is higher during the day and lower at night	

Layer F2: The electron density is about 10^{11}–10^{12} e/m^3 ranging from 200 to 500 km, and its peak height fluctuates between 300 and 400 km.

Layer F1 and F2 are collectively referred to as layer F, which is responsible for the major transmission of radio waves as skywave propagation.

The region above the altitude HMF2 (F2-layer electron-concentration peak), where the maximum electron density of layer F2 is detected thousands of kilometers above, is, in its entirety, referred to as top ionosphere or upper ionosphere, and the region below altitude HMF2 is referred to as bottom ionosphere or lower ionosphere. In the upper ionosphere region, electron density decreases gradually. The electron density is about 1012–10^{10} e/m^3 at 1,000 km altitude; and the density is about 10^{11}–10^8 e/m^3 between 2,000 and 3,000 km altitude. The electron density of the upper ionosphere demonstrates substantial variations with the changing seasons, as well as with day and night. The height of each layer, extent of electron density, and basic features of ionosphere are shown in Table 5.1.

5.1.2 Ionospheric Models

Contemporarily, the prevalent ionospheric empirical and semi-empirical models include IRI (International Reference Ionosphere), CRI (China Reference Ionosphere), NeQuick model, and Klobuchar model. Among them, IRI is regarded as the most effective ionospheric empirical model used extensively at present; CRI incorporates China's observation data of ionosphere into the IRI model to make it more suitable for China; NeQuick-G model, being a rapid computation method for electron density, is adopted by the Galileo system. The IRI, CRI, and NeQuick models are introduced in this section.

1. IRI:

IRI, the global standard empirical ionospheric model, has been developed by the IRI work group in conformity with extensive ground observation data and accumulated ionospheric research achievements, and is funded and endorsed conjointly by COSPAR (Committee on Space Research) and URSI (International Union of Radio Science).

IRI divides the ionosphere into six regions, including upper region, layer F_2, layer F_1, middle region, peak region, bottom region in layer E, and layer D; the profile structure is shown in Fig. 5.2. This model integrates several atmospheric parameter models pertaining to vertical inclinometer, incoherent scatter radars, satellite data, data of sounding rockets, etc.; it introduces a monthly average parameter of solar activity and geomagnetic activity index along with determining the monthly average states of ionosphere ranging between 80 and 2,000 km above a specific location at a specific time under certain inobtrusive geomagnetic conditions, such as location-specific electron density, electron temperature, ion temperature, ion composition, and electron content. IRI has two sets of standby coefficients, namely, CCIR coefficient established in 1967 and URSI coefficient in 1989, that can project the global ionosphere morphology more effectively.

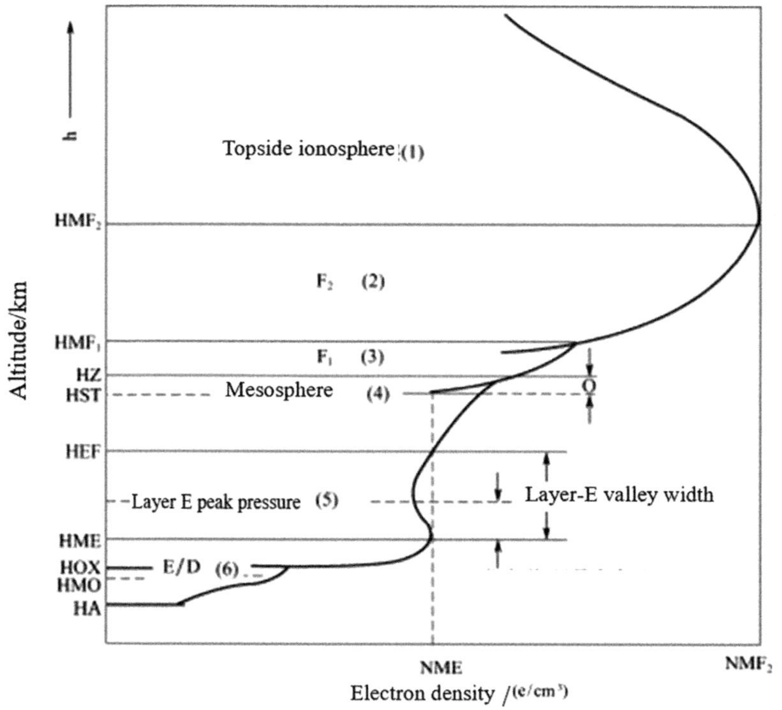

Fig. 5.2 IRI [2] ①. Width from the bottom of Layer E to the point with equal concentration of maximum electrons of Layer E

IRI global model was established in 1957, and it gets updated annually, releasing a new version each year in accordance with the contemporary research achievements and observation data. The latest version of the same is IRI2016. Since the year 2000, the IRI work group has been incorporating the GIM (global ionosphere model) data and the observation results of several other space radio detection technologies into the IRI model, to enhance its precision level.

2. CRI:

Ionospheric observation data of China is seldom used for the development of the IRI model, and, hence, the said model is hardly accurate for China. China Research Institute of Radio Propagation uses the IRI model only as a reference, and employs the ionospheric observation data obtained in China to formulate the CRI, thereby enhancing the precision of the model for China efficiently.

Compared to the IRI model, the CRI model exhibits the following improvements:

(1) Adopts f_0F_2 and M (3000) F_2 data provided in "Ionospheric Prediction for Asia and Oceania" method.
(2) Collects the occurrence time of layer F_1 in China, and recommend to consider that layer F_1 exists in all seasons in China.
(3) The height of the maximum electron density in layer E is 115 km.

3. NeQuick Model:

① Explanations for the English abbreviations are as follows:

HZ: Upper boundary of mesosphere.
HST: The height to which the electron-concentration profile in the F_1 layer is pushed down to be equal to the maximum electron concentration in the E-layer;
HEF: upper boundary of layer E.
HDX: Specific height of layer D and bottom of layer E.
HA: The starting height of the ionosphere.
HMF2, HMF1, HME, and HMED refer to the heights corresponding to the maximum electron concentration of layers F_2, F_1, E, and D.
ME, HMF2: The maximum electron concentration of layers E and F_2.

The NeQuick model is an innovative ionospheric model that is jointly researched by ARPL-ICTP, Trieste, and IGAM, University of Graz. This model can provide spatial and temporal distribution of electron density of the ionosphere completely from bottom to top, and maximum output height of the model is up to 20,000 km. Basic output parameters of the model are longitude and latitude, height, time, solar activity level (solar radiation flux or sunspot number, SSN) of a station, electron density under certain given conditions, and total electron content of the vertical ionosphere above this station. Additionally, the NeQuick model also offers a computation method to calculate the electron content of the ionosphere in the vertical path, between any two points.

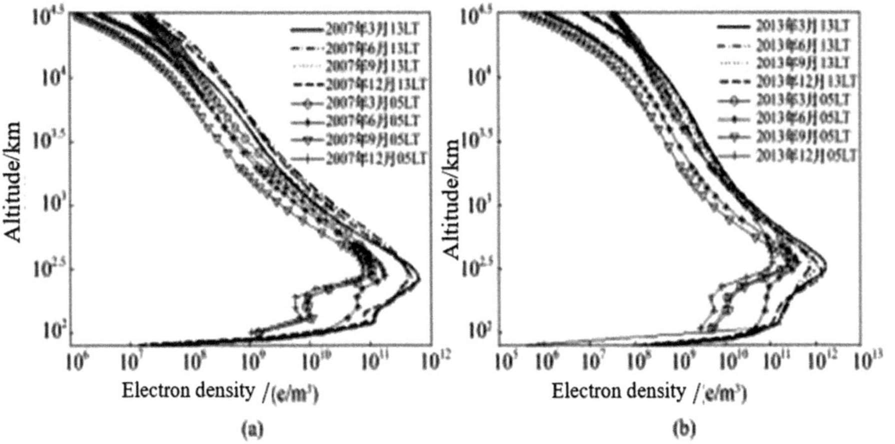

Fig. 5.3 Profile of typical electron density height in low solar activity year (2007, (**a**)) and high solar activity year (2013, (**b**))

Using the NeQuick model, Fig. 5.3 illustrates the electron density profile of two typical moments (05:00LT① and 13:00LT) in spring, summer, autumn, and winter for a low solar activity year (2007) and a high solar activity year (2013) in Beijing at mid-latitude.

5.1.3 *Ionospheric Total Electron Content (TEC)*

Total Electron Content (TEC) is the integration result of electron density "Ne" along the radio wave propagation path "s," that is, the number of electrons contained in the column per unit area across the bottom of the whole ionosphere along the radio wave propagation path, and forms an important parameter of the ionosphere:

$$TEC = \int_{s} N_e(s)\, ds \tag{5.1}$$

In the formula: The unit of TEC is TECU, 1TECU $= 1 \times 10^{16}$ e/m^2. When a radio wave signal travels along the zenith direction, the electron content on this path is called as Vertical Total Electron Content (VTEC).

① LT is the abbreviation of Local Time.

According to existing studies, Bottom Electron Content (BEC) accounts for about 20–40% of VTEC, while Upper Electron Content accounts for about 60–80% of the same. The electron content at different altitudes is listed in Table 5.2.

Table 5.2 Percentage of electron content at different altitudes

Altitude scope	Percentage of VTEC/(%)
Bottom electron content	20–40
Upper electron content	60–80
Below 1,000 km	80–90
Above 1,000 km	10–20
Above 2,000 km	5–10

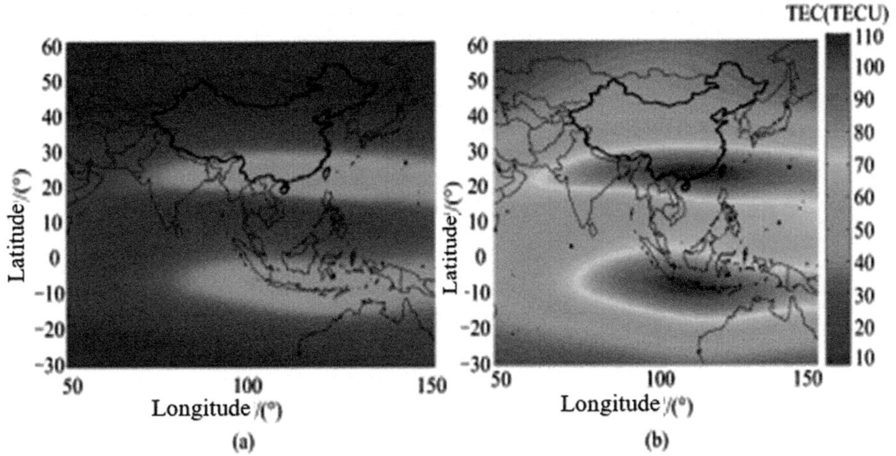

Fig. 5.4 VTEC distribution in low and high solar activity year in China and surrounding regions. **a** 2007 (low); **b** 2013 (high)

Calculation of VTEC's distribution in China and surrounding regions at 13:00LT in low solar activity year (2007) and high solar activity year (2013) using the NeQuick model is shown in Fig. 5.4. According to the figure, VTEC is higher at both the sides of the equator, and lower above the equator, showing a double-hump structure.

We can adopt GNSS's (Global Navigation Satellite Systems) observation data to inverse the VTEC outcome. Figure 5.5 demonstrates the inversion result of VTEC for 3 consecutive days for a low (2007) and high solar activity year (2013) in Beijing. According to the figure, VTEC recorded in the year of high solar activity is much larger than in low solar activity year, and is significantly characterized by diurnal variation, that is, the VTEC is lowest between 04:00 and 05:00LT, and largest between 14:00 and 16:00LT.

Using the NeQuick Model, Fig. 5.6 demonstrates the change of TEC with elevation angle of satellite-ground links in the morning (05:00LT) and afternoon (13:00LT) in summer (June) and autumn (September) for low solar activity year (2007) and high solar activity year (2013), at an altitude of 1,000 km.

According to the above analysis:

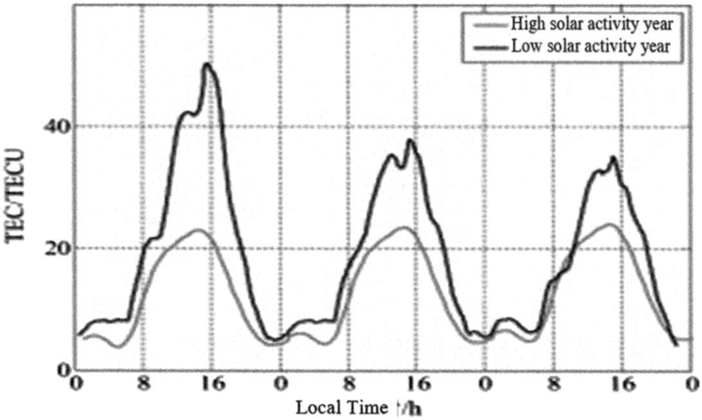

Fig. 5.5 Daily variation characteristics of VTEC

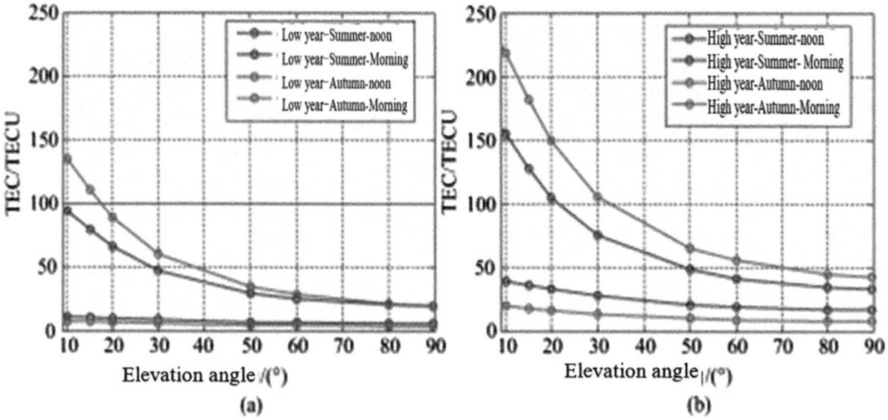

Fig. 5.6 Curve of TEC changing with elevation angle in low year (2007, (**a**)) and high year (2013, (**b**))

(1) TEC is relative to solar activity level, and TEC in high solar activity year is larger than in low solar activity year.
(2) TEC is significantly characterized by the regions evaluated. In view of the geographical latitude, TEC at low latitudes is larger than that at middle and high latitudes under identical conditions.
(3) TEC exhibits evident daily variation or diurnal characteristics, and it is higher during the day than at night.
(4) TEC is relative to radio wave propagation path. The lower the elevation angle, the larger the TEC.

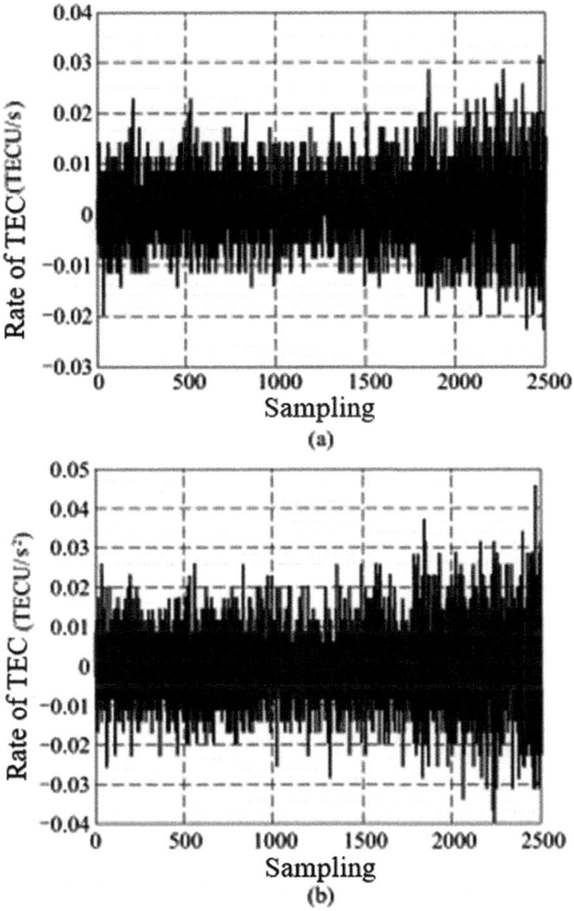

Fig. 5.7 Change rate (**a**) and secondary change rate (**b**) of TEC

Furthermore, TEC is also affected by seasonal change. It is higher in spring and autumn seasons and lowest in summer. Occasionally, huge irregularities are also found in the variation manner of TEC.

Using GNSS measurement data, Fig. 5.7 provides Beijing TEC change rate at 08:00LT (Fig. 5.7a) and secondary change rate (Fig. 5.7b), in which the primary change rate is about 0.01TEC/s. The maximum change rate observed by the satellite crossing aurora zone with orbital altitude of 22,000 km is about 0.7TEC/s.

5.1.4 Ionospheric Irregularities

In addition to the typical layering structure, as shown in Fig. 5.1, certain "cloud-shaped" or "wave-shaped" structures are at times seen forming in the ionosphere

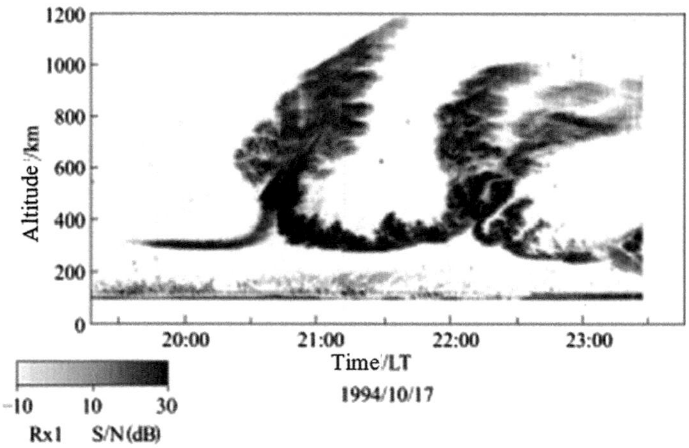

Fig. 5.8 Ionospheric irregularities observed by incoherent scatter radar [3]

due to the irregularity in electron density, known as ionospheric irregularities. The electron density of ionospheric irregularities may be lower than or higher than the average electron density of the ionosphere, and the maximum relative change of electron density is usually only a few percent, but can reach up to 10–100% in extreme cases.

The common ionospheric irregularities primarily entail traveling ionospheric disturbance (TID), sporadic E-layer (Es-layer), and Spread-F. The spatial scale range of ionospheric irregularities shows a massive variation ranging from thousands of kilometers to dozens of centimeters, triggering a rapid and random fluctuation of phase, amplitude, and polarization of signals crossing ionosphere, which can result in severe scintillation fading of satellite signals. Figure 5.8 provides the ionospheric irregularities observed by an incoherent scatter radar.

Sporadic E-layer (Layer Es) is a prevalent irregularity found in layer E. Figure 5.9 provides a profile of electron density of the two events of sporadic E-layer observed by incoherent scatter radar. According to the figure, electron density of sporadic E-layer is much larger than that of the background altitude, and, at times, even larger than that of layer F.

Spread-F is a common irregularity found in layer F. Figure 5.10 illustrates the ionogram given by an ionosonde of a vertical inclinometer in normal and spread-F events. As seen in the figure, F echoes clearly when the ionosphere is in normal state, so that the echoes of ordinary wave (O-wave) and extraordinary wave (X-wave) can be seen more evidently, and all appear in the form of a line; when spread-F occurs, echo of layer F is not a line but is dispersed. This is attributable to the scattering effect of ionospheric irregularities on the signals, such that the echo pulse reflected by layer F is wider than the transmission pulse and could be even 10 times wider in extreme cases.

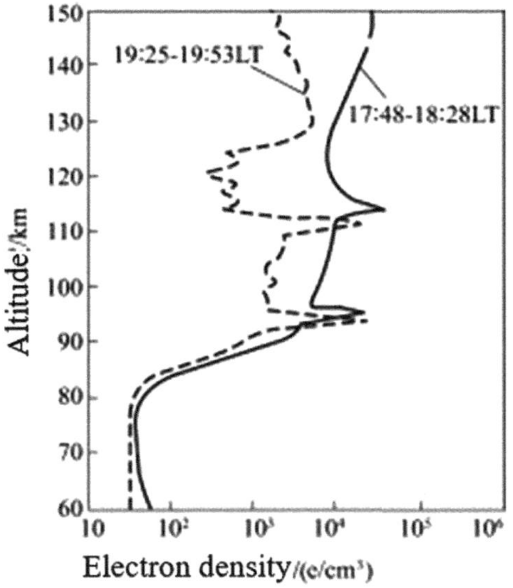

Fig. 5.9 Sporadic E-layer event observed by Arecibo incoherent scatter radar [4]

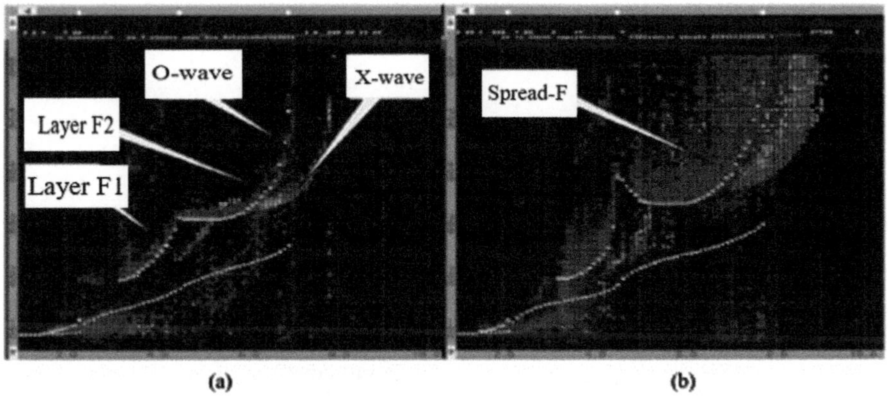

Fig. 5.10 Ionospheric tracing observed by ionosonde [5]. **a** Ionospheric tracing in normal condition; **b** ionospheric tracing of Spread-F

5.2 Background Ionospheric Propagation Effects

Background ionosphere refers to the silent ionospheric state, devoid of the ionospheric disturbance. Ionosphere is a kind of dispersive plasma, in which the refractive index varies with frequencies, such that the radio waves propagating at different frequencies travel at different speeds resulting in the widening of the transmission pulse. At the same time, the layering attribute of the ionosphere also induces refraction and multipath propagation effect of the radio waves. The magnetic field of the earth makes the ionosphere anisotropic, and the incident wave is divided into ordinary wave (O-wave) and extraordinary wave (X-wave), bringing about Faraday's rotation effect. Moreover, the refractive index of the ionosphere is less than 1, and, hence, its group velocity tends to be smaller than the velocity of light when a radio wave propagates in the ionosphere, and the phase lead and delay effect are put into motion if the phase velocity is larger than the velocity of light. Consequently, the background ionosphere could trigger propagation effects, such as absorption, phase shift, delay, dispersion, polarization rotation, refraction, and multipath, thereby influencing the performance of a satellite system in numerous ways.

5.2.1 Ionospheric Refractive Index

When a radio wave propagates in the ionosphere, the refractive index is described using the Appleton-Hartree equation:

$$n^2 = 1 - \frac{X}{1 - iZ - \left(\frac{Y_T^2}{2(1-X-iZ)}\right) \mp \sqrt{Y_L^2 + \frac{Y_T^4}{4(1-X-iZ)^2}}} \tag{5.2}$$

In which:

$$Y_T = \frac{f_H}{f}\sin\theta \cong \frac{\mu_0 H_0 |e|}{2\pi m}\frac{\sin\theta}{f}$$
$$Z = \frac{v_e}{f}$$
$$X = \frac{f_p^2}{f^2} = \frac{N_e e^2}{4\pi^2 \varepsilon_0 m f^2} \cong \frac{80.6}{f^2}N_e$$
$$Y_L = \frac{f_H}{f}\cos\theta \cong \frac{\mu_0 H_0 |e|}{2\pi m}\frac{\cos\theta}{f}$$

In the formula: f_p refers to plasma frequency (Hz) and f refers to frequency (Hz); N_e denotes electron density (e/m^3); e refers to electronic charge ($1.60217733 \times 10^{-19}$ C); m represents electron mass ($9.1093897 \times 10^{-31}$ kg); ε_0 refers to free-space dielectric constant (8.854×10^{-12} F/m); θ refers to angle of geomagnetic field to propagation wave (rad); H_0 refers to strength of geomagnetic field (A/m); f_H refers to electron cyclotron frequency (Hz); μ_0 refers to free-space magnetic conductivity (1.256637×10^{-7} H/m); and v_e denotes electron collision frequency (Hz).

When there is applied magnetic field, the plasma is magnetized. The magnetized plasma is a birefringent medium, which is also known as anisotropic medium. In

Formula (5.2), "+" and "L" represent ordinary wave (O-wave) and extraordinary wave (X-wave).

According to the Appleton-Hartree formula, the ionosphere possesses the following characteristics due to the geomagnetic field influence:

(1) Dispersion: Refractive index is the function of frequency, causing the group velocity of a radio wave to be unequal to phase velocity.
(2) Absorption: Refractive index is a complex number with a real and an imaginary part. The imaginary part corresponds to the absorption of the ionosphere, and absorption is always dissipated; hence, the energy is converted into heat energy by collision and remains conserved during the propagation process.
(3) Birefringence: Due to the existence of a geomagnetic field and free electrons, the refractive index attains different values, and this characteristic indicates that a radio wave has two possible propagation paths, each of which possesses a different phase velocity and group velocity.
(4) Anisotropy: Refractive index also serves as the function of background magnetic field. Under the influence of background magnetic field, refractive indices exhibit anisotropy because they vary differently in the different propagation directions.

In the above VHF frequency case, $X << 1$, Formula (5.2) can be simplified to

$$n_p \approx 1 - \frac{X}{2} \approx 1 - 40.3\frac{N_e}{f^2} \tag{5.3}$$

The refractive index of phase propagation n_p and of group propagation n_g of a radio wave in the ionosphere satisfies $n_g = n_p + f\frac{\partial n_p}{\partial f}$, and thus,

$$n_g \approx \frac{1}{n_p} = 1 + 40.3\frac{N_e}{f^2} \tag{5.4}$$

5.2.2 *Group Delay*

The ionosphere behaves as a dispersive non-linear medium. Radio waves propagating at different frequencies experience diverse refractive indices, such that the propagation velocity of wave packet energy (group velocity v_g) is smaller than the velocity of light, but the propagation velocity of carrier phase (phase velocity v_p) is greater than the velocity of light. Phase velocity and group velocity can be represented as follows, respectively:

$$v_p = \frac{c}{n_p} \tag{5.5}$$

$$v_g = \frac{c}{n_g} \tag{5.6}$$

In the formula: n_p refers to phase refractive index, as shown in Formula (5.3); n_g refers to group refractive index, as shown in Formula (5.4).

When radio waves of a satellite system propagate within the ionosphere, the waves propagate at group velocity v_g, and the geometric distance ρ from the satellite to ground is as follows:

$$\rho = \int_{\Delta t} v_g dt = \int_{\Delta t} c\left(1 - 40.3\frac{N_e}{f^2}\right) dt = c \cdot \Delta t - \frac{40.3}{f^2}\int_{s'} N_e\, ds = P - \frac{40.3}{f^2}\int_{s'} N_e\, ds \tag{5.7}$$

In the formula: f refers to frequency (Hz), c refers to velocity of light (2.99792458 $\times$ 10^8 m/s), and P refers to pseudorange (m), which is the product of propagation time Δt and velocity of light c.

The distance delay of a radio wave signal caused by the ionospheric parameters is the difference between pseudorange P and geometric distance ρ:

$$\Delta l = \frac{40.3}{f^2}\int_{s'} N_e\, ds = \frac{40.3}{f^2} TEC \tag{5.8}$$

Correspondingly, the propagation delay is

$$\Delta \tau = \frac{\Delta l}{c} = \frac{40.3}{cf^2}\int_{s'} N_e\, ds = \frac{40.3}{cf^2} TEC \tag{5.9}$$

In the formula: $\Delta\tau$ is the time delay (s), f is frequency (Hz), and TEC is in e/m^2.

When the TEC value in the ionosphere is different, the time delay $\Delta\tau$ varies with respect to frequency f, as shown in Fig. 5.11.

For the 1.6 GHz radio signal, when TEC changes from 1TECU (10^{16} e/m^2) to 1000TECU (10^{19} e/m^2), time delay $\Delta\tau$ ranges between 0.5 and 500 ns. Figure 5.12 portrays the distribution of annual average value of percentage of global day-time delay exceeding 20 ns in high solar activity year (sunspot SSN = 140), with frequency of 1.6 GHz.

5.2.3 *Phase Advance*

Due to the phase refractive index of the ionosphere being less than 1, the radio wave propagates with phase velocity v_p, which is greater than the velocity of light, to generate a phase advance Δl_φ (m).

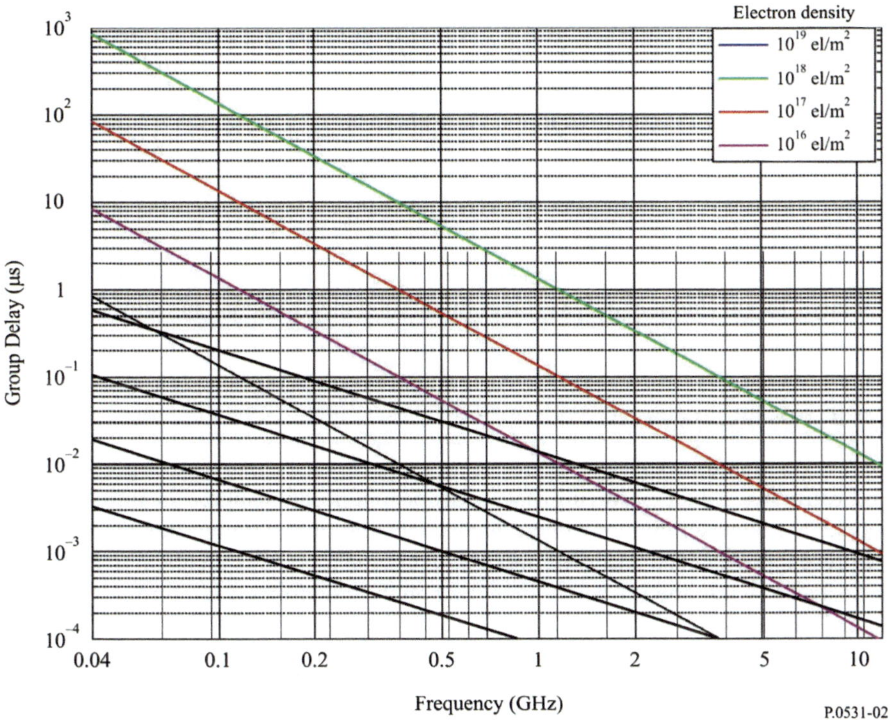

Fig. 5.11 Time delay varying with frequency and TEC [6]

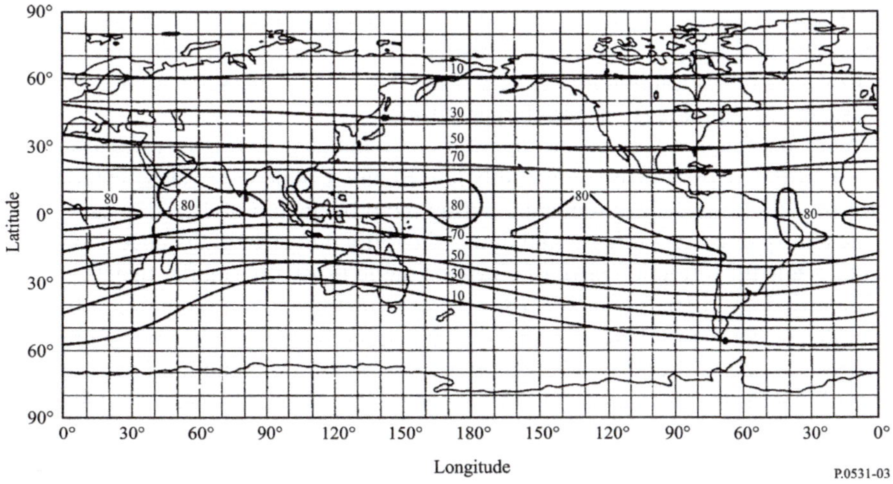

Fig. 5.12 Annual average percentage of day-time delay exceeding 20 ns in high solar activity year

$$\Delta l_{\varphi} = -\frac{40.3}{f^2}\int\limits_{s'} N_e \, ds \tag{5.10}$$

Phase advance Δl_{φ} (m) due to the ionosphere is

$$\Delta\varphi = \frac{2\pi}{\lambda}\left|\Delta l_{\varphi}\right| = 2\pi\frac{40.3}{cf}TEC \tag{5.11}$$

In the formula: f refers to frequency (Hz), λ refers to wavelength (m), c refers to velocity of light (m/s), and TEC refers to the total electron content (e/m^2).

5.2.4 Refraction

The refractive effect occurring due to tropospheric atmosphere is introduced in Sect. 2.1. Likewise, the ionosphere is also capable of generating this effect due to the presence of ionized particles, and its calculation method is identical to that given in Sect. 2.1, while the refractive index is simply transformed into group refractive index in this case.

When a radio signal passes through the ionosphere, the propagation path bends and the group velocity of radio wave appears smaller than the free-space velocity of light, resulting in the delay of signal and the refraction error. Unlike the troposphere, the ionosphere is a dispersive medium, and the refraction varies with the signals of different frequencies. Figure 5.13 illustrates the refraction error of ionospheric distance and elevation angle refraction in UHF frequency band (500 MHz) in Beijing. According to this figure, the error of distance at a low elevation angle can measure up to a scale of several hundred meters, while the error of elevation angle can be up to a scale of milliradians, or even larger at times.

When the frequency of a radio signal is at S band (2–4 GHz), the refraction error caused by the ionosphere is significantly low. Figure 5.14 demonstrates the refraction error of distance and elevation angle during the propagation of 2.5 GHz radio wave in the ionosphere. According to the figure, the refraction error of distance is of meter grade, while the refraction error of elevation angle is usually smaller than 0.5 mrad.

Figures 5.15, 5.16, 5.17 and 5.18 provide refraction errors (including refraction error due to the troposphere and ionosphere) caused by typical elevation angles (3°, 5°, 10°, 15°, 20°, 30°, 60°) in low solar activity year (2007) and high solar activity year (2013) at 01:00LT (a) and 13:00LT (b) at UHF frequency band (500 MHz) in Beijing, and the target altitude is 1,000 km.

Based on the above analysis, the refraction of the ionosphere exhibits the following characteristics:

(1) The ionosphere is a dispersive medium; hence, its refraction effect is relative to frequency. The lower the signal frequency, the larger is the refraction error. For UHF frequency band, the refraction error of distance due to low elevation

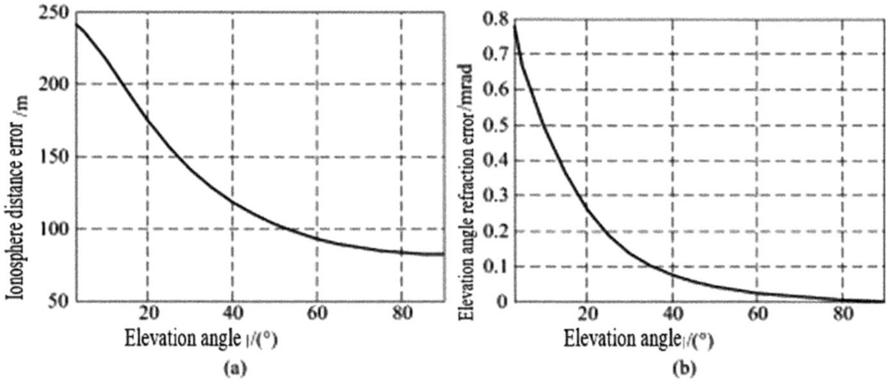

Fig. 5.13 Refraction error of ionospheric distance and elevation angle refraction in UHF frequency band

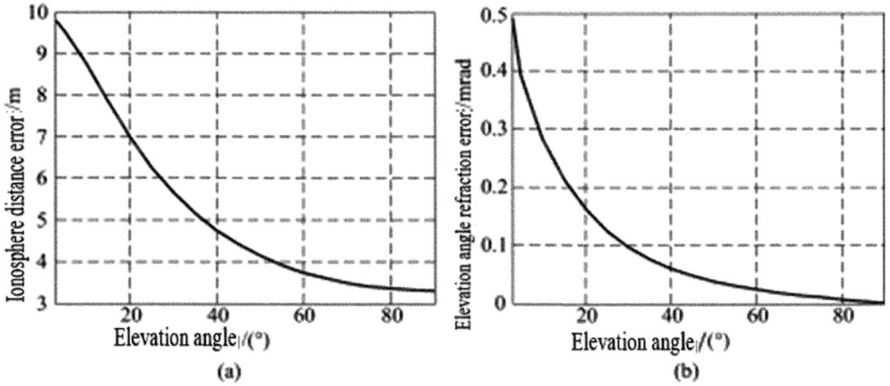

Fig. 5.14 Refraction error of distance and elevation angle at S frequency band

angle can measure up to a scale of hundreds of meters, and error of elevation angle can be up to milliradian scale.

(2) The refraction error due to ionosphere is relative to solar activity and geographic position and is highlighted by apparent diurnal variation.

Essentially, the refraction error of distance exhibits identical variation features as TEC and is significantly highlighted by seasons and regions under the influence of solar activity.

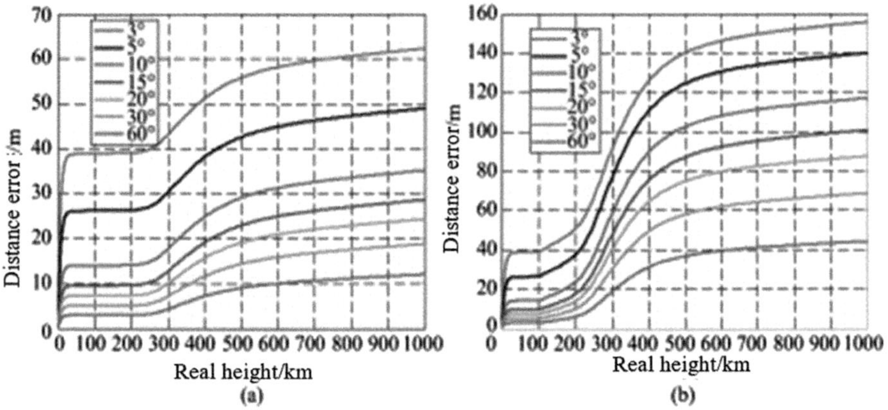

Fig. 5.15 Error of distance at 01:00LT (**a**) and 13:00LT (**b**) in March for low solar activity year (2007)

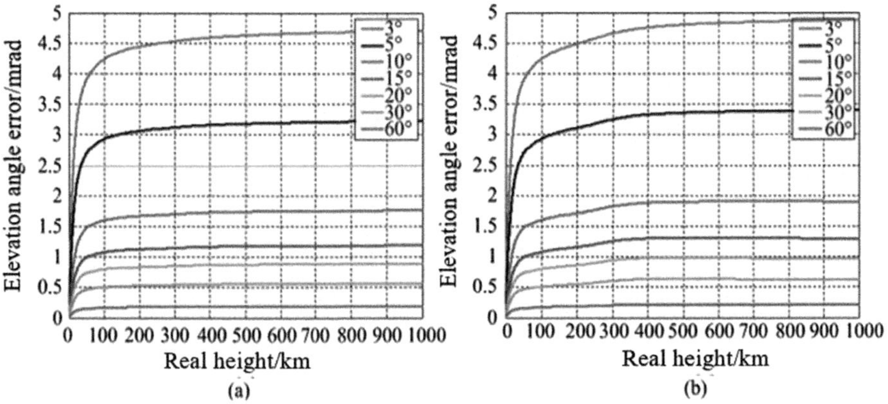

Fig. 5.16 Error of elevation angle at 01:00LT (**a**) and 13:00LT (**b**) in March for low solar activity year (2007)

5.2.5 *Dispersion*

Since the refractive index is frequency dependent, propagation delay also tends to be frequency dependent for the radio wave propagation in the ionosphere, due to which, different frequency components of the pulse spectrum possess different propagation delays. Therefore, the receiving pulse signal gets distorted owing to such dispersion effect, and can severely compromise the performance of a broadband system; for example, the imaging resolution of a wide-band imaging radar reduces considerably due to such a distortion.

Delay dispersion typically refers to change of rate of frequency corresponding to delay, that is,

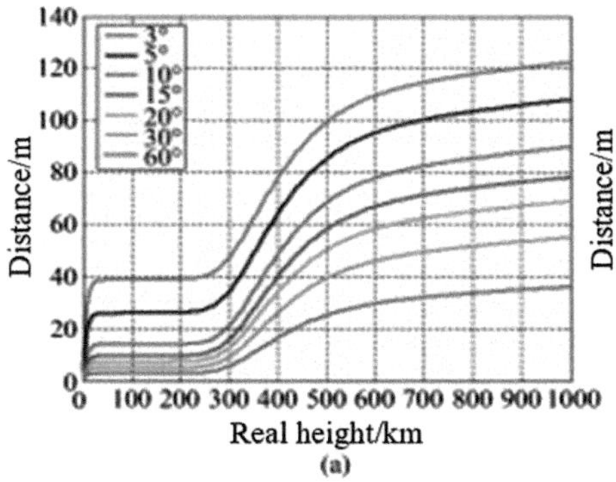

Fig. 5.17 Error of distance at 01:00LT (**a**) and 13:00LT (**b**) in March for high solar activity year (2013)

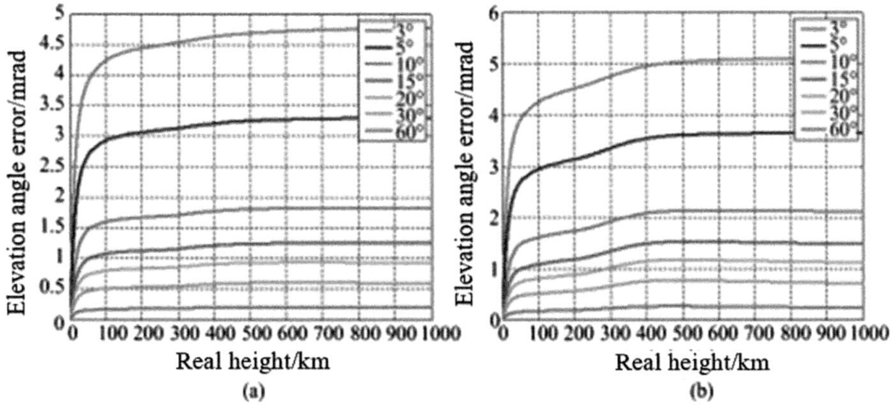

Fig. 5.18 Error of elevation angle at 01:00LT (**a**) and 13:00LT (**b**) in March for high solar activity year (2013)

$$\frac{dt}{df} = -\frac{2.68 \times 10^{-7}}{f^3} TEC \tag{5.12}$$

In the formula: $\frac{dt}{df}$ refers to delay dispersion (s/Hz), f refers to frequency (Hz), and TEC refers to total electron content (e/m^2).

When signal frequency is 0.1–10 GHz, pulse widths are recorded to be 0.01, 0.1, 1, and 10 μs; when TEC is 50TECU (50×10^{16} e/m^2), delay dispersion due to the ionosphere is as shown in Fig. 5.19. Thus, when the pulse width of signal τ is 1 μs, delay dispersion is 0.017 μs in case of 200 MHz of frequency and delay dispersion is 0.00062 μs in case of 600 MHz of frequency. It can be observed that delay dispersion effect produces a substantial impact on the broadband transmission system operating at VHF and UHF bands.

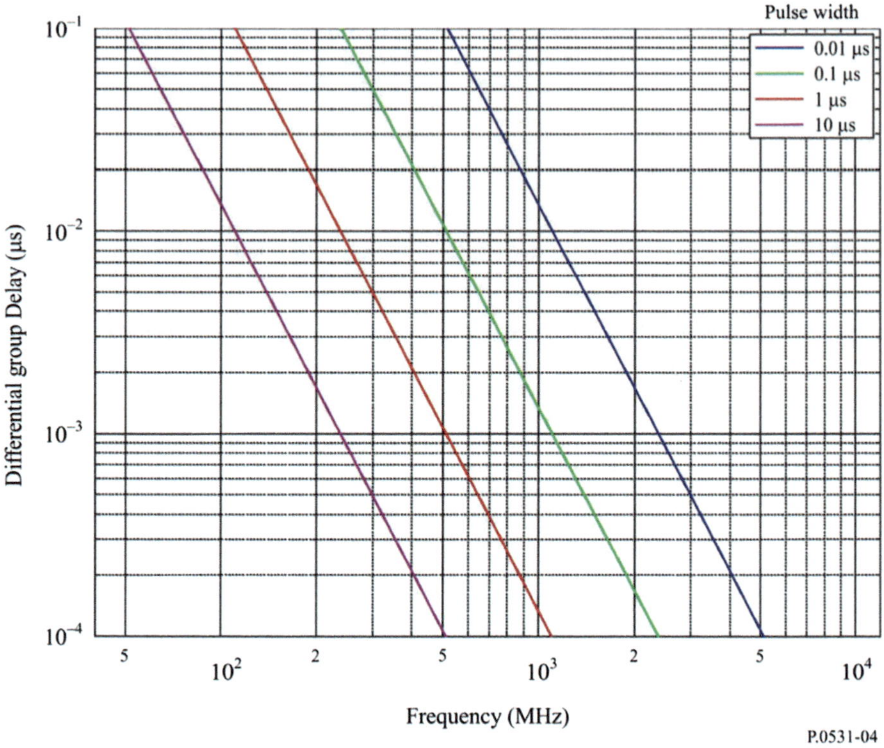

Fig. 5.19 Delay dispersion effect on signals with different pulse widths [6]

Phase dispersion $\frac{d\varphi}{df}$ (rad/Hz) due to ionosphere is as follows:

$$\frac{d\varphi}{df} = -2\pi \frac{40.3}{cf^2} TEC = -\frac{8.44 \times 10^{-7}}{f^2} TEC \tag{5.13}$$

In the formula: f refers to frequency (Hz) and c refers to velocity of light (m/s). Second-order phase dispersion $\frac{d^2\varphi}{df^2}$ (rad/Hz) is as follows:

$$\frac{d^2\varphi}{df^2} = d\left(\frac{-8.44}{f^{-2}} \times 10^{-7}\right)/df = -\frac{16.88}{f^3} \times 10^{-7} \tag{5.14}$$

5.2.6 Faraday Rotation

Existence of earth's geomagnetic field makes the ionosphere an anisotropic medium. After a linearly polarized radio wave arrives into the ionosphere, it is split into ordinary wave (O-wave) and extraordinary wave (X-wave) possessing different wave vectors and phase velocities. These two waves are then elliptically polarized in opposite polarization rotation direction. The higher the frequency, the more circularly polarized the two elliptical polarization components. Since the two components possess different propagation speeds, a linearly polarized wave is reconstructed after propagating through a certain distance in the ionosphere, and it is rotated in the vector direction by a certain angle with respect to wave vector of the incident waves, which is called as Faraday rotation effect or magneto-optic Faraday effect.

For VHF frequency band and above, the Faraday rotation angle Ω (rad) due to the ionosphere is

$$\Omega = \frac{K_1}{f^2}\int N_e B_L \, dl \tag{5.15}$$

In the formula: constant $K_1 = 2.36 \times 10^4$, f refers to frequency (Hz), N_e refers to electron density (m^{-3}), B_L refers to strength of geomagnetic induction (T) parallel to propagation direction, B_{LF} refers to average geomagnetic field strength (T) parallel to propagation direction, and is also called as Faraday geomagnetic factor.

When receiving the radio waves from a geostationary satellite at middle and low latitudes, the value of B_{LF} often is

$$B_{LF} = B(h)\cos\varphi(h)|_{h=h_m=420\,\text{km}} \tag{5.16}$$

In the formula: h_m refers to average altitude of ionosphere, being 420 km generally; $B(h)$ refers to geomagnetic induction strength at h_m altitude in the propagation path; and $\varphi(h)$ refers to an angle of ray propagation path at h_m altitude to local field direction.

The Faraday rotation angle Ω can be represented as

$$\Omega = \frac{2.36 \times 10^4 B_{LF}}{f^2}\int N_e \, dl \tag{5.17}$$

Therefore, Faraday rotation angle Ω is relative with radio frequency f, average geomagnetic field strength along integration path BLF, and total electron content along propagation path TEC.

For a radio wave in 0.1–10 GHz frequency band, when TEC changes between 1 TECU (10^{16} e/m^2) and 1000TECU (10^{19} e/m^2), the Faraday rotation angle Ω due to the presence of ionosphere is shown in Fig. 5.20.

For a linear antenna, the cross-polarization discrimination (XPD) due to Faraday rotation can be calculated using the following formula:

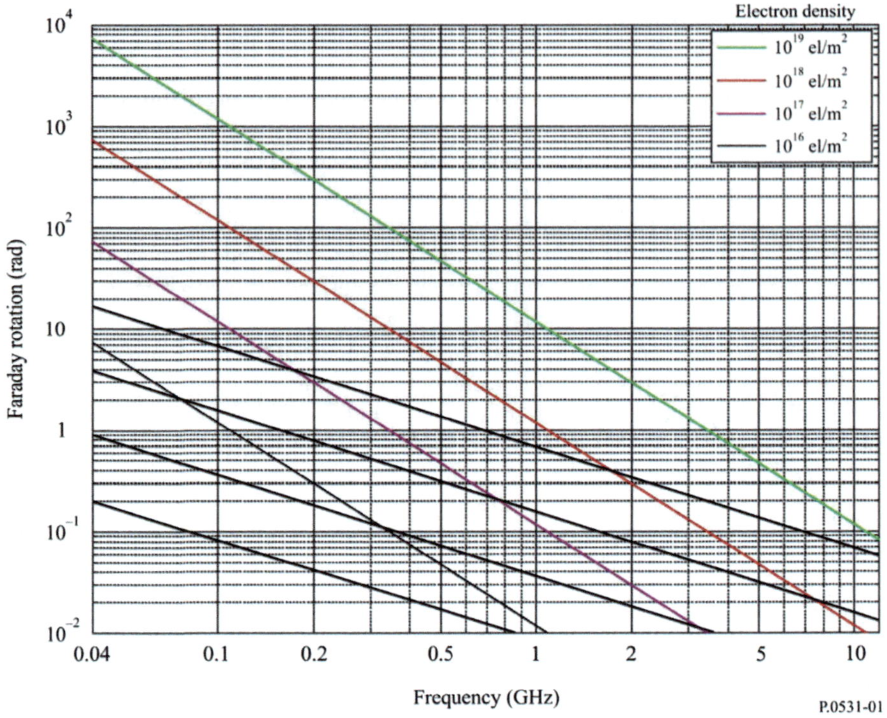

Fig. 5.20 Relationship between Faraday rotation, frequency, and TEC [2]

$$XPD = -20\log(\tan\Omega) \tag{5.18}$$

Polarization loss P_L (dB) can be represented as

$$P_L = -20\log(\cos\Omega) \tag{5.19}$$

5.2.7 Doppler Effect

The change of frequency for one-way Doppler shift due to the ionosphere is

$$\Delta f_d = \frac{1}{2\pi}\frac{d(\Delta\varphi)}{dt} = -\frac{40.3}{cf}\frac{dTEC}{dt} \tag{5.20}$$

In the formula: Δf_d refers to Doppler frequency shift (Hz), f refers to frequency (Hz), c refers to velocity of light (m/s), and TEC refers to total electron content (e/m^2).

Excluding the ionosphere, the Doppler frequency shift for propagation signals of satellite-ground links is only dependent on the component of speed of transmitter (or receiver) along the ray's direction.

When the ionosphere is in an inactive state, $\frac{dTEC}{dt}$ is very small (with an average of 0.2TECU/s, see Sect. 5.2.1), but during intense ionospheric activity, TEC changes rapidly with a sharp increase in the Doppler frequency shift. For a GNSS system, when Doppler frequency shift increases (for example, $\Delta f_d > 1$ Hz), the phase-locked loop of GNSS receiver fails to locate the phase of GNSS signal, which results in locking failure.

5.2.8 Absorption

For 30 MHz radio waves and above, the absorption occurring in the ionosphere is directly proportional to (sec i)/f^2, where "i" refers to the zenith angle of propagation path and "f" refers to the frequency considered. At equatorial and low-latitude regions, with respect to 70 MHz and above systems, the absorption taking place is too minute to be observed.

According to the trial and measurement, the absorption of 30 MHz radio signal is generally around 0.2–0.5 dB at middle-latitude regions when it vertically passes through the ionosphere a single time. The ionospheric absorption increases significantly and can reach up to 5 dB during the solar flare period. Absorption enhancement phenomena generally occur at high-latitude regions, such as auroral absorption and polar-cap absorption. They appear randomly and vary in durations, and their spatial and temporal distribution characteristic is determined by the observation test.

At polar regions, the neutral atmosphere of region D and region E is ionized due to the presence of sedimented high-energy electrons bringing about a surge in electron density. When a radio signal passes through the ionosphere, the electron movement is induced by the field vector of the wave, and each electron collides with other particles leading to attenuation of the radio wave amplitude, which is regarded as auroral absorption of the ionosphere. The auroral absorption typically occurs between 10° and 20° positioned centrally on the latitude having the highest probability of occurrence of visible auroras, which is manifested as a series of sporadic absorption with relatively short duration (from several minutes to hours, with an average of 30 min), and shows a smooth rapid rise and gradual descend for absorption at night.

Polar cap absorption occurs at a region having geomagnetic latitude larger than 64°, with relatively small probability of occurrence. This absorption always occurs in peak solar activity year, 10 times to 12 times annually, lasting several days each time. The occurrence of polar-cap absorption is always accompanied by discrete solar activity and can be observed at sunward polar-cap regions. Such an absorption naturally shows a reduced activity at night.

The auroral absorption and polar-cap absorption are inversely proportional to the square of radio frequency involved. Table 5.3 provides the typical auroral absorption of 127 MHz radio wave in accordance with ITU-RP.531 Recommendation.

Table 5.3 Auroral absorption of 127 MHz radio wave [6]

Time percentage/(%)	Elevation angle	
	5°	20°
0.1	1.5	2.9
1	0.9	1.7
2	0.7	1.4
5	0.6	1.1
20	0.2	0.4

In summary, being a dispersive medium, the ionosphere affects the radio signals at different frequencies differently, with various parameters coming into play, such as time delay, phase advance, refraction, dispersion, Faraday rotation, Doppler frequency shift, and absorption. Furthermore, the angle of arrival of the receiving signal also varies to a certain extent. Table 5.4 provides the analysis of ionospheric propagation effects for 0.1–10 GHz radio waves, and it corresponds to the single-way path with 30° elevation angle, and TEC value of about 100TECU (10^{18} e/m^2).

Table 5.4 Analysis of ionospheric propagation [7]

Effect	Frequency relation	0.1 GHz	0.25 GHz	0.5 GHz	1 GHz	3 GHz	10 GHz
Time delay (μs)	f^{-2}	25	4	1	0.25	0.028	0.0025
Refraction	f^{-2}	<1°	<0.16°	<2.4′	<0.6′	<0.4″	<0.36″
Change of angle of arrival	f^{-2}	20′	3.2′	48″	12″	1.33″	0.12″
Faraday rotation	f^{-2}	30 cycles	4.8 cycles	1.2 cycles	108°	12°	1.1°
Absorption (aurora or polar cap) (dB)	f^{-2}	5	0.8	0.2	0.05	0.006	0.0005
Absorption (middle latitude) (dB)	f^{-2}	<1	<0.16	<0.04	<0.01	<0.001	<0.0001
Time-delay dispersion	f^{-3}	400 ps/kHz	26 ps/kHz	3.2 ps/kHz	0.4 ps/kHz	14.5 ts/kHz	0.4 ts/kHz

5.3 Ionospheric Scintillation

5.3.1 Ionospheric Scintillation Characteristics

1. Overview:

Inhomogeneity of electron density in the ionosphere can induce random fluctuations in dielectric constants and refractive indices. When a radio wave propagates in such an unsystematic medium, the propagation path and delay also vary randomly, and the signals arriving from different paths undergo interference at the receiving antenna, eventually effectuating rapid fluctuations of amplitude and phase of the receiving signal, which is also known as ionospheric scintillation. Figure 5.21 illustrates the ionospheric scintillation on GPS $L1_1$ (1575.42 MHz) and L_2 (1227.6 MHz) signal. It can be observed that the radio signal appears as a stable curve before the occurrence of ionospheric scintillation, while the receiving signal fluctuates rapidly during the scintillation event (21:00–22:30 UT).

Ionospheric scintillation is caused by ionospheric irregularities, which are the small-scale disturbances generally occurring in layer F at an altitude of 250–400 km. These irregularities emerge from the instability of signals at the receiver terminal

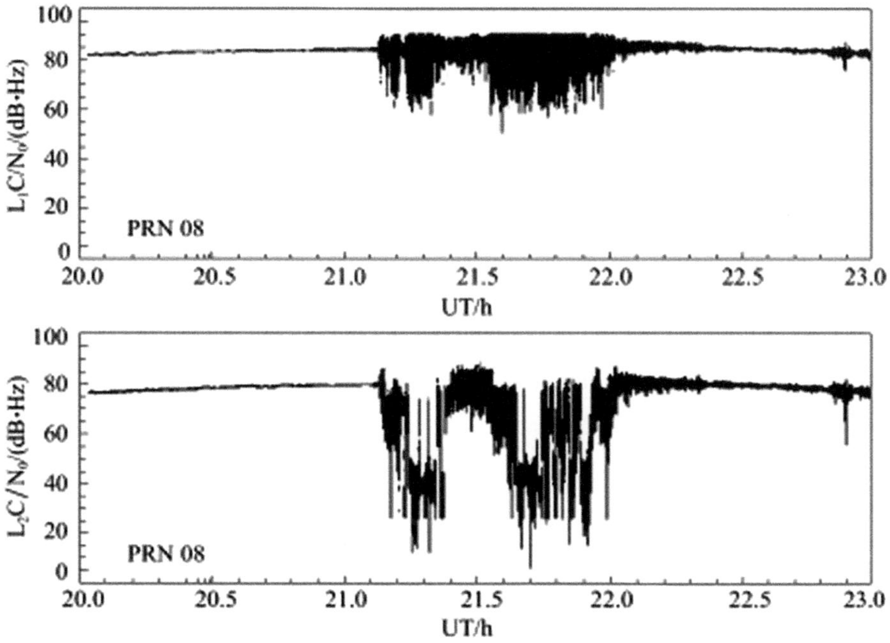

Fig. 5.21 Example of ionospheric scintillation

through forward scattering and diffraction of radio waves, triggering rapid fluctuations of amplitude, phase, and angle of arrival, and also changing their correlation at the same time.

The ionospheric scintillation may affect the radio signals in wider frequency bands, especially signals below 3 GHz.

In fact, the occurrence of ionospheric scintillation is observed in all the trials until the frequency reaches up to 10 GHz. Ionospheric scintillation greatly compromises the quality of satellite communication, and a strong scintillation event could even cause critical interruption of satellite signals in some frequency bands.

2. Index of ionospheric scintillation:

The index of ionospheric scintillation S_4, used to describe the strength of ionospheric scintillation, refers to the ratio of standard deviation of receiving power to average receiving power within a certain interval:

$$S_4 = \left(\frac{\langle I^2 \rangle - \langle I \rangle^2}{\langle I \rangle^2} \right)^{\frac{1}{2}} \tag{5.21}$$

In the formula: "I" refers to signal intensity and symbol < > refers to statistical average.

According to the amplitude of scintillation index S_4, the scintillation is divided into no scintillation ($S_4 < 0.1$), weak scintillation ($0.1 \le S_4 < 0.3$), moderate scintillation ($0.3 \le S_4 < 0.6$), and strong scintillation ($S_4 \ge 0.6$).

3. Relationship between scintillation index and elevation angle:

For weak scintillation and moderate scintillation, the scintillation index S_4 is directly proportional to the cosecant value of elevation angle e of earth-space links; when elevation angle of earth-space links is $20° \le a \le 90°$, $S_4 \propto \csc(a)$; when elevation angle of earth-space links is a <20°, $S_4 \propto$ cscα(i), the range of α is $1/2 \le \alpha < 1$.

4. Relationship between scintillation index and longitude:

Index of ionospheric scintillation S_4 is season and longitude dependent, and can be represented as

$$S_4 \propto \exp\left(\frac{\beta}{W}\right) \tag{5.22}$$

In the formula: angle β refers to the angle of magnetic longitude of earth-space links and ionosphere pierce point (IPP) to magnetic longitude of the sun at sunset, $H = 350$ km; weighted value W is the geographic position that is station dependent, and it varies greatly for different regions and for each day in a year.

Figure 5.22 illustrates the change of W for an observed day of a year in three different stations, including Kwajalein (9°5/N, 167°20′E), Hong Kong (22°16′N,

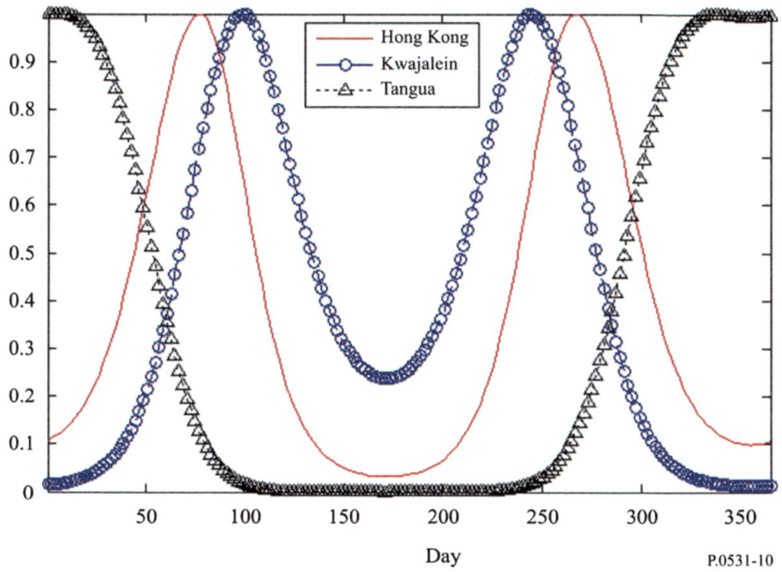

Fig. 5.22 Seasonal weighting function of stations at different latitudes [6]

114°9′E), and Tangua (22°44′S, 42°43′W). The curve of W for Hong Kong can be used as reference data for low-latitude regions in China. This curve can be approximately represented as

$$W = 0.71 + 0.29\sin[2(t - 34) \times \frac{\pi}{180}] \tag{5.23}$$

In the formula: t refers to observed day of the year; average of W is 0.71, it is strong at the spring and autumn equinox, and weak at summer and winter solstice.

5. Relationship between scintillation index and frequency:

Scintillation index S_4 is radio frequency dependent. The index S_4 for weak and moderate scintillation indisputably relates to f^{-v} and can be around 1.5 for most of the frequencies. For strong scintillation, the factor observed by trial is reduced due to the influence of repetitive strong scattering. When S_4 is approaching 1, the amplitude abides by Rayleigh distribution.

Figure 5.23 provides the indexes of 150 MHz (dotted line) and 400 MHz (dashed line) ionospheric scintillation observed at Kiruna (a), Lulea (b), and Kokkola (c) station by receiving the beacon signal of LEO satellite Tsykada during ionospheric disturbance on October 30, 2003. It can be observed that the scintillation index S_4 may exceed 1 and is up to 1.5 in case of occurrence of strong scintillation, and even much larger at times.

6. Spatial and temporal distributions of ionospheric scintillation:

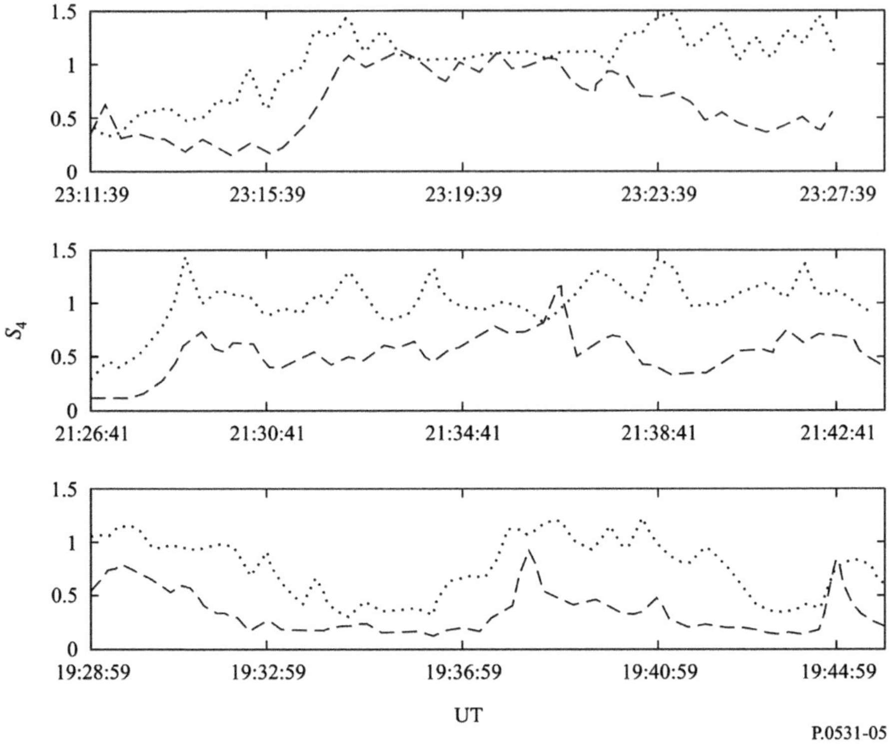

Fig. 5.23 VHF and UHF scintillation index S_4 observed at **a** Kiruna Station, **b** Lulea Station, and **c** Kokkola Station

According to long-term observation and statistical analysis, ionospheric scintillation is characterized by

(1) The occurrence time of ionospheric scintillation is largely affected by external factors, such as solar activity. By and large, the scintillation occurs 1 h after sunset or late in the evening, often between 19:00 and 06:00.
(2) Ionospheric scintillation in high solar activity year is stronger and more frequent than in low solar activity year and is often much stronger in spring (March to April) and autumn (September to October) of high solar activity year.
(3) In terms of the geographic position, scintillation always occurs near the equator (within $\pm 20°$ of geomagnetic latitude), followed by the high-latitude regions, such as aurora zone and polar-cap zone. Ionospheric scintillation only occurs under abnormalities, such as geomagnetic storm, and is less likely to occur or is quite weak at middle-latitude regions.

7. Ionospheric scintillation model:

The global ionospheric scintillation model (GISM) has been developed by the European Space Agency and adopted by ITU-R. This model consists of two parts. The first part comprises the NeQuick background ionosphere model, in which the electron density in the ionosphere is first calculated, and then the fluctuation of electron density is obtained by using the solution of ray difference equation. The second part contains Multi-Phase Screen (MPS) technology. Herein, radio wave propagation path can be determined by obtaining electron density and dividing the disturbed propagation medium into several continuous phase screens with clear statistical features using MPS technology, and then finally deriving the index of ionospheric scintillation of earth-space links by adopting the solution of corresponding radio wave propagation equation.

For a satellite system, ionospheric scintillation is relative to satellite and position of ground station, solar activity, geomagnetic activity, local time, and frequency. GISM can predict the scintillation index, amplitude fading depth, and root-mean-square value of phase and angle deviation. The default value of chief internal parameters of the model is as follows:

(1) Slope of intensity spectrum, $k = 3$.
(2) Average scale of irregularities, $L_0 = 500$ km.
(3) Standard deviation of electron density fluctuation, $\sigma_{Ne} = 0.2$.

5.3.2 *Scintillation Amplitude Distribution and Spectrum Characteristics*

In a certain ionospheric scintillation event, the signal amplitude distribution can be described approximately by Nakagami probability density function. The relative signal density I (with respect to frequency carrying power of average power), that is, normalized average power abides by Nakagami distribution, and its probability density function is expressed as

$$p(I) = \frac{m^m}{\Gamma(m)} I^{m-1} \exp(-mI) \tag{5.24}$$

In the formula: coefficient m of Nakagami distribution is scintillation index S_4 dependent:

$$m = 1/S_4^2 \tag{5.25}$$

The probability of receiving the signal with density below the given threshold can be represented as

$$P(I) = \int_0^I p(x)\,dx = \frac{\Gamma(m, mI)}{\Gamma(m)} \tag{5.26}$$

In the formula: $\Gamma(m)$ is complete gamma function and $\Gamma(m, mI)$ is incomplete gamma function, which can be defined as

$$\Gamma(m) = \int_0^\infty x^{m-1} \exp(-x)dx \tag{5.27}$$

$$\begin{aligned}\Gamma(m, mI) &= \int_0^I (mx)^{m-1} \exp(-mx)d(mx) \\ &= \int_0^{mI} y^{m-1} \exp(-y)dy\end{aligned} \tag{5.28}$$

The percentage of receiving the signal with density above or below the given threshold in an ionospheric scintillation event can be calculated in accordance with Formula (5.26). For example, the time probability of signal x lower than average value is P (10 − X/10) and time probability of signal y higher than average value is 1 – P (10y/10).

According to extensive observation results, the power spectrum density of ionospheric scintillation varies between f^{-6} and f^{-1}, and the typical power spectrum density is shown in Fig. 5.24.

In the figure, A refers to relative power spectrum density of radio signal 30 min prior to the scintillation event; B refers to relative power spectrum density of radio signal at the time of occurrence of the scintillation event; and C, D, E, and F refer to relative power spectrum density of signal 1 h, 2 h, 3 h, and 4 h after the scintillation event.

For practical application, the power spectrum density varies with frequency and can be taken as f^{-3}, if there is no effective measurement result available.

5.3.3 Statistics Characteristics of Ionospheric Scintillation

While designing a satellite communication system, it is crucial to consider statistical characteristics of scintillation effects together with the influence of a single scintillation event on the system.

1. Scintillation peak-peak fluctuation value:

The relationship between scintillation index and scintillation peak-peak fluctuation value P_{fluc} (dB) can be approximately represented as

$$P_{\text{fluc}}\,(\text{dB}) = 27.5 S_4^{1.26} \tag{5.29}$$

The calculation result of Formula (5.29) is shown in Fig. 5.25.

The cumulative distribution $P(I)$ of signal intensity I can be obtained by using cumulative distribution of signal peak-peak fluctuation value.

In the formula:

$$P(I) = \sum_{i=0}^{n} f_i P_i(I) \tag{5.30}$$

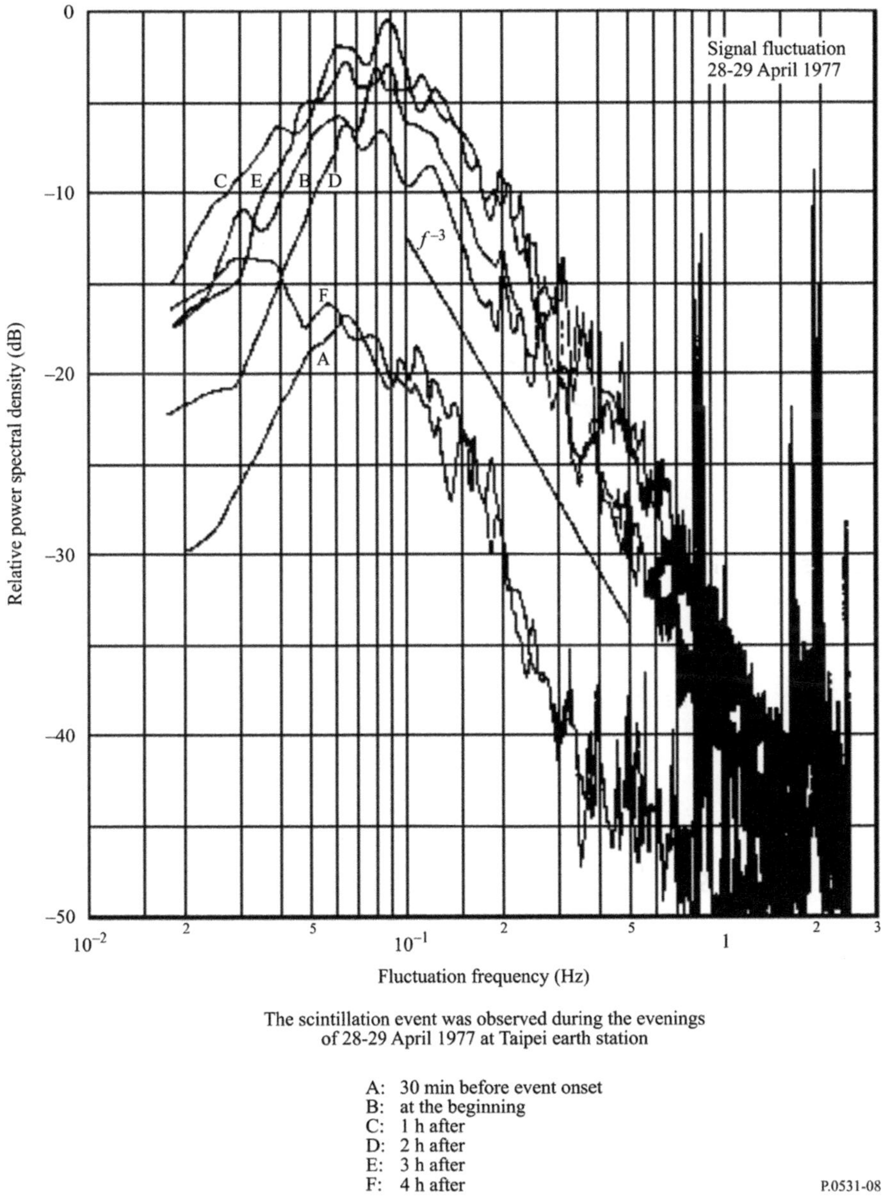

Fig. 5.24 4 GHz signal power spectrum observed at night on April 28 and 29, 1977 in Taipei, China [2]

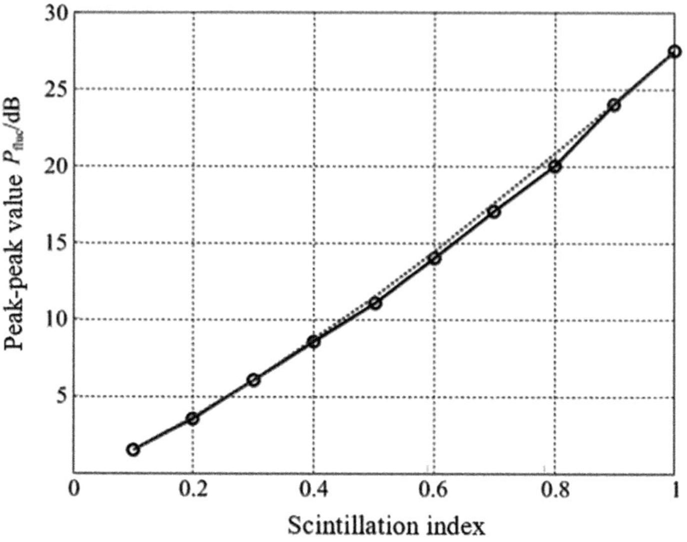

Fig. 5.25 Relationship between scintillation index and signal peak-peak fluctuation value

$$
\begin{gathered}
f_0 = F(\xi < \xi_1) \\
f_i = F(\xi_i \le \xi < \xi_{i+1}) \quad (i = 1, 2, \cdots\cdots n-1) \\
f_n = F(\xi \ge \xi_n) \\
P_i(I) = \Gamma(m_i, m_i I)/\Gamma(m_i) \\
m_i = 1/S_{4i}^2 \\
S_{40} = \left[\frac{1}{27.5}\cdot\frac{\xi_1}{2}\right]^{1/1.26} \\
S_{4i} = \left[\frac{1}{27.5}\cdot\frac{\xi_i+\xi_{i+1}}{2}\right]^{1/1.26} \quad (i = 1, 2, \cdots\cdots n-1) \\
S_{4n} = \left[\frac{1}{27.5}\cdot\frac{\xi_{n-1}+3\xi_n}{4}\right]^{1/1.26}
\end{gathered}
$$

In the formula: ξ_1, ξ_n refer to minimum and maximum values of signal peak-peak value and n refers to number of intervals that the user is interested in.

The statistical relationship between peak-peak value of ionospheric scintillation and sunspot number is derived in accordance with 4 GHz ionospheric scintillation event (Table 5.5) observed at Hong Kong, Taipei, Bahrain, and Longovilo Stations from 1970 to 1980, as shown in Fig. 5.26.

2. Evaluation of scintillation effect:

The annual statistics of peak-peak fluctuation values of ionospheric scintillation are obtained using 4 GHz ionospheric scintillation events (Table 5.6) observed in Hong Kong and Taipei from 1975 to 1980, as shown in Fig. 5.27, in which, sunspot number is the 12-month sliding mean, solid-line curve P represents the measured result of

Table 5.5 Statistics of ionospheric scintillation event [6]

Event	Time	Station	Number of events
A	1975–1976	Hong Kong, Bahrain	15
B	1974	Longovilo	1
C	1976–1977	Taipei	2
D	1970–1971	12 stations	+50
E	1977–1978	Hong Kong	12
F	1978–1979	Hong Kong	10
G	1979–1980	Hong Kong	6

satellite angle with east elevation angle of 20°, and dash curve *I* represents measured result of satellite signal with west elevation angle of 30°.

The scintillation effect of a satellite system is analyzed by the following methods, as shown in Fig. 5.27:

(1) The peak-peak fluctuation value P_{fluc} of 4 GHz signal scintillation with required time probability obtained from Fig. 5.27.
(2) The peak-peak fluctuation value with frequency f is obtained by multiplying peak-peak fluctuation value of 4 GHz signal scintillation by $(f/4)^{-1.5}$.
(3) P_{fluc} varies with geographic position and occurrence time per day and can be evaluated qualitatively using Fig. 5.23.
(4) When calculating sanction of links, the signal attenuation Lp can be obtained by $Lp = P_{\text{f1uc}}/\sqrt{2}$.

Figure 5.28 depicts an example of long-term cumulative distribution of 4 GHz satellite signal density.

3. Scintillation fading depth:

The scintillation fading depth $A_1(X)$ can be represented as

$$A_I(x) = \begin{cases} 0 & S_4 < 0.05 \\ -10\log I(x) & S_4 \geq 0.05 \end{cases} \tag{5.31}$$

In the formula: $A_1(x)$ refers to fading depth with $x\%$ time probability exceeded (dB) and $I(x)$ denotes the relative receiving level with $x\%$ time probability exceeded, which can be defined as

$$I(x) = \frac{P_r}{\overline{P_r}} = \frac{E_r^2}{\overline{E_r^2}} \tag{5.32}$$

In the formula: Pr refers to receiving power level (dBw), $\overline{P_r}$ refers to average receiving power (dBw), Er denotes field intensity of receiving point, and $\overline{E_r^2}$ refers to mean square value of field intensity.

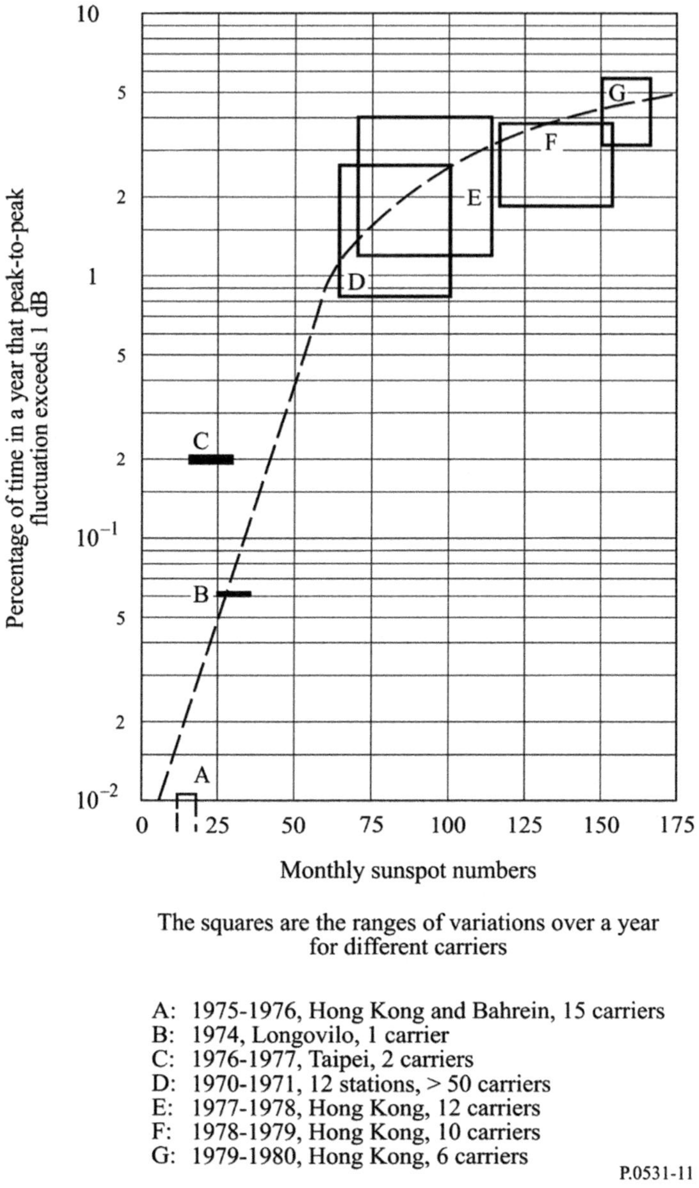

Fig. 5.26 Relationship between peak-peak fluctuation value of ionospheric scintillation and sunspot number [6]

Table 5.6 Statistics of ionospheric scintillation event [6]

Curve	Time	SSN scope	Station
I1, P1	1975–1976, 03	10–15	Hong Kong
I2, P2	1976–1977, 06	12–26	Taipei
I3, P3	1977–1978, 03	20–70	Hong Kong
I4, P4	1977–1978, 10	44–110	Hong Kong
I5, P5	1978–1979, 11	110–160	Hong Kong
I6, P6	1979–1980, 06	153–165	Hong Kong

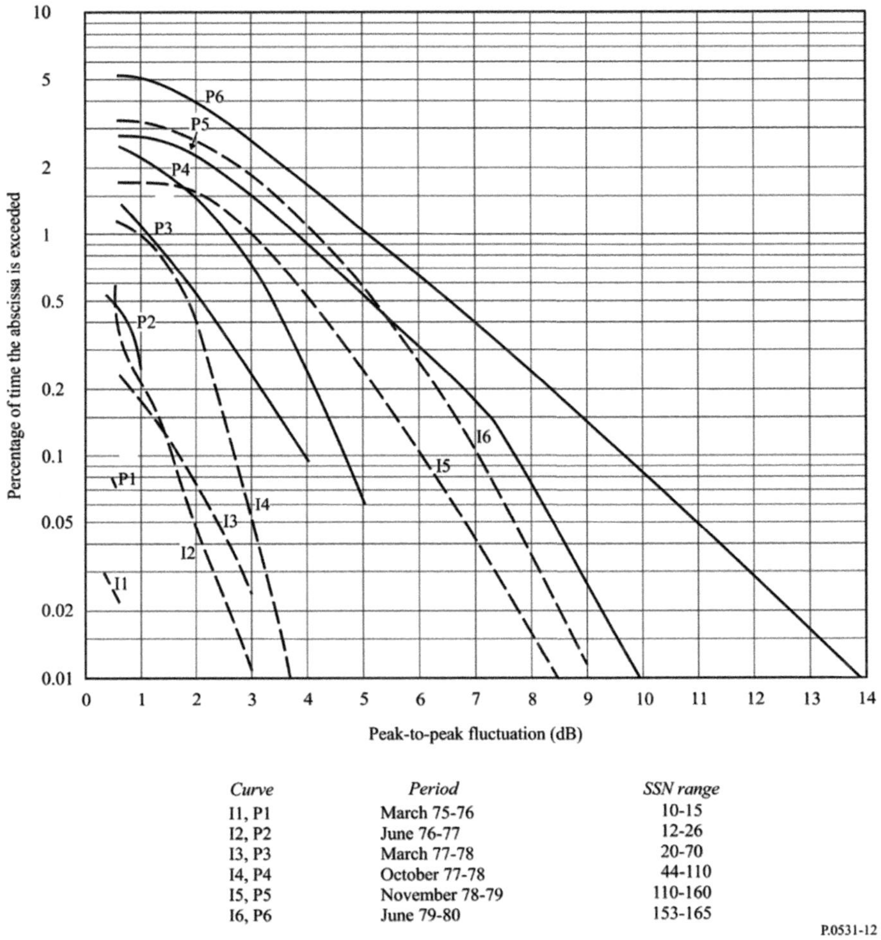

Fig. 5.27 Annual statistics of peak-peak fluctuation value of 4 GHz signal ionospheric scintillation [6]

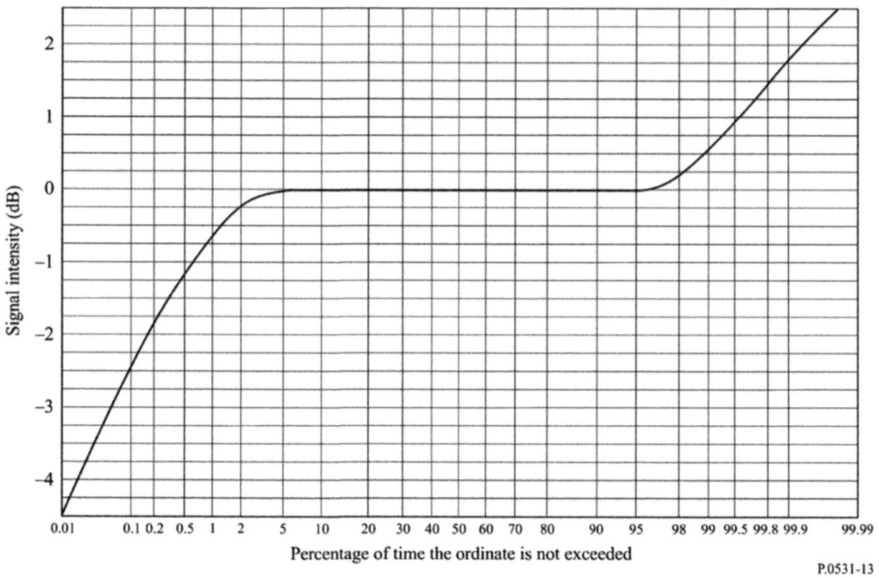

Fig. 5.28 Example of long-term cumulative statistics of satellite signal density (40 GHz, 20° of elevation angle) [6]

Figure 5.29 depicts the global distribution of scintillation fading depth of low-frequency band signal (1.5 GHz) in a high solar activity and a low solar activity year, based on which we can deduce that the fading depth has the following characteristics:

(1) The scintillation fading in high solar activity year is fairly larger than that in low solar activity year.

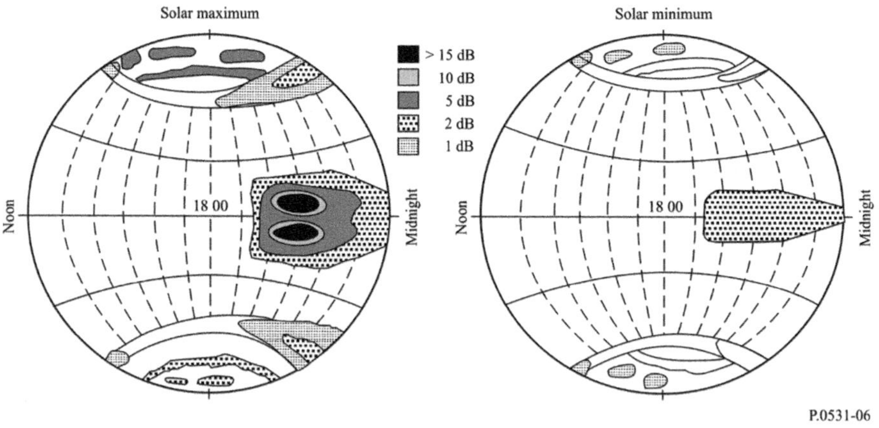

Fig. 5.29 Scintillation fading depth distribution in high solar activity year and low solar activity year [6]

(2) The scintillation fading usually occurs at 19:00LT in the evening, lasting till 06:59LT in the morning, sometimes.
(3) The fading frequently occurs near the equator (within ±20° of geomagnetic latitude), followed by the high-latitude regions, such as aurora zones and polar-cap zones, while the scintillation is either minimal or very less likely to occur in the middle-latitude regions.

For 3 GHz frequency system and above, the effect of rainfall on the signals should be considered (refer to Chap. 3) in addition to ionospheric scintillation. In fact, attenuation due to ionospheric scintillation and rain attenuation refer to the losses of two different mechanisms, but such losses may occur at the same time at the equator and its surrounding regions during a high solar activity year. For example, in Djutiluhar in Indonesia, the trial shows the annual probability of simultaneity of scintillation fading and rain attenuation of 4 GHz signal to be about 0.06%. The signal is not depolarized if only scintillation fading takes place, and the fluctuation amplitude of signal is relatively small if only rain attenuation occurs; however, severe signal fluctuation can occur in the channels of cross-polarization if both the phenomena transpire simultaneously.

5.3.4 Ionospheric Scintillation Characteristics in China Region

The southern low-latitude regions in China are situated near the equator, where the occurrence of ionosphere scintillation is high; hence, China's satellite system is severely impacted from this phenomenon. Therefore, China Research Institute of Radio Propagation has been working on ionospheric scintillation observation and research for a long time by utilizing radio observation stations and grids. It analyzes the data from Haikou, Guangzhou, and Kunming Stations in China to obtain a comprehensive understanding of the characteristics of ionospheric scintillation in China's low-altitude regions.

1. Ionospheric scintillation characteristics in China's low-altitude regions:

(1) The ionospheric scintillation phenomenon is largely solar activity dependent. The occurrence probability of ionospheric scintillation per day can be up to 80% and above in a high solar activity year, and the probability is higher in spring and autumn and lower in summer and winter. Figure 5.30 displays the statistical result of correlation between ionospheric scintillation and solar activity index observed at Haikou Station from December 2008 to June 2014, wherein the red line refers to solar activity index F10.7, blue line refers to solar activity index SSN, and bar graph refers to occurrence probability of ionospheric scintillation. The Y-coordinate on the left refers to the occurrence probability of ionospheric scintillation, y-coordinate on the right refers to the solar activity index SSN, and x-coordinate refers to the observed month. It can be observed that the occurrence

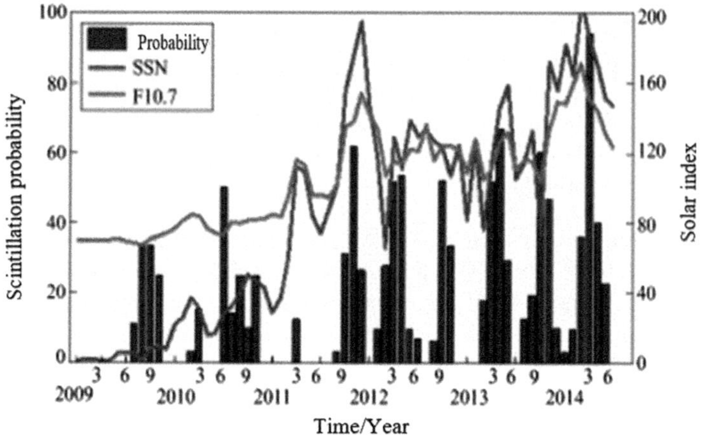

Fig. 5.30 Correlation analysis of solar activity index and ionospheric scintillation [8]

probability of ionospheric scintillation exhibits a good and positive correlation with solar activity index SSN (or F107).

(2) The scintillation predominantly occurs in the evening after sunset, is concentrated between 21:00LT at night and 02:00LT the next morning, and its occurrence is less likely during the day.

Figure 5.31 displays the correlation between occurrence of ionospheric scintillation events during high occurrence seasons (spring (February to October) in 2005 and autumn (August to October) in 2004) at the local time in China's Hainan Province.

(3) The ionospheric scintillation in China always occurs in the regions south of geographic latitude 28°N (geomagnetic latitude is about 18°N). Figure 5.32 depicts the ionospheric scintillation observed by the GNSS signal. In the figure, blue area refers to the region without scintillation; gray area refers to weak scintillation ($0.1 \le S_4 < 0.3$ for weak scintillation), green area refers to moderate scintillation ($0.3 \le S_4 < 0.6$ for moderate scintillation), and red area refers to strong scintillation ($S_4 \ge 0.6$ for strong scintillation).

(4) At low-latitude regions, the ionospheric scintillation is more likely during the geomagnetically inactive period, and less likely in the geomagnetically active period.

Figure 5.33 depicts the occurrence statistics of ionospheric scintillation during different geomagnetic activity periods, signifying that ionospheric scintillation predominantly occurs during the geomagnetically inactive period and is inhibited to a large extent in case of a strong geomagnetic activity period.

2. Statistical characteristics of ionospheric scintillation:

In terms of a satellite system design, Fig. 5.34 illustrates the scintillation fading depth with %P (95, 99, 99.5, 99.9, 99.95, and 99.99%) time probability exceeded for

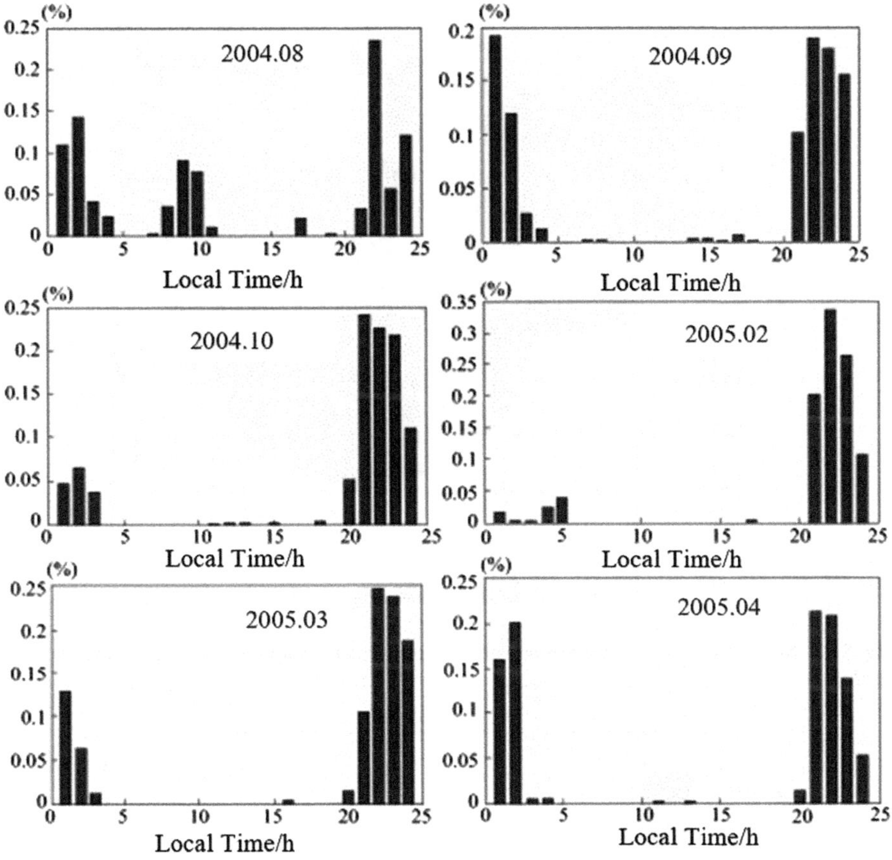

Fig. 5.31 Local time variation of scintillation occurrence probability in Hainan [8]

0.1–10 GHz signals in China's low-latitude regions, wherein x-coordinate refers to the signal frequency (GHz) and y-coordinate refers to the fading depth (dB).

In the middle-latitude regions, ionospheric scintillation seldomly occurs, and scintillation fading depth is relatively small; hence, its influence can be overlooked for 1 GHz signals and above. Similar to Figs. 5.34 and 5.35 provides the scintillation fading depth with *p*% (99, 99.5, 99.8, and 99.9%) time probability exceeded for 0.1–1 GHz signals in the middle-latitude regions.

References

1. Xiong N, Tang C, Li X (1999) An introduction to ionosphere physics. Wuhan University Press, Wuhan
2. Peinan J, Zhongzhi Z (2007) Radar environment and characteristics of radio wave propagation. Publishing House of Electronics Industry, Beijing
3. Guo-jun W (2007) Observation characteristics of ionospheric irregularity in low latitude (Hainan) region. PhD dissertation, Chinese Academy of Sciences
4. Tan H (2004) Statistical characteristics and simulation of the mid-latitude ionospheric sporadic E-layer. PhD dissertation, Chinese Academy of Sciences
5. Li JH (2006) Study on ionospheric scintillation and parameter inversion of irregular body. Master dissertation, Xidian University
6. REC. ITU-R P.531-12 (2013) Ionospheric propagation data and prediction methods required for the design of satellite services and systems
7. Zhengwen X (2005) Study on the influence of ionosphere on satellite signal propagation and performance. PhD dissertation, Xidian University
8. Sheng D (2014) Study on ionospheric scintillation characteristics and current reporting method in low latitude region of China. Master Dissertation, Electronic Science Research Institute
9. Ippolito LJ (1999) Propagation effects handbook for satellite systems design, 5th edn

Chapter 6
Radio Noise and Interference Coordination

In a satellite communication system, the satellite or ground station can receive numerous external noise signals along with the main radio signals, which further generates a considerable impact on the quality of satellite communication. In particular, the external noise signals tend to be much larger than the interferences in the receiver owing to the usage of a low-noise amplifier, and thus we must consider the influence of the various external noises when designing the communication links of an earth-space satellite. Furthermore, the ground station of the satellite interferes with the ground service system on the same frequency. In system design, the coexistence analysis must be explored to define the coordination distance from the ground station. When defining such distance, it is imperative to consider the interference caused by beyond-line-of-sight (BLOS) propagation of rainfall satellite and atmospheric duct, in addition to the line-of-sight propagation of the interference signal.

6.1 Definition and Classification of Radio Noise Signals

Radio noise, which is deemed as a time-varying electromagnetic phenomenon with a radio-frequency component that evidently does not transmit any valuable information, and can, in fact, superimpose or get merged with useful information is divided into internal and external noise. Internal noise emerges from the receiving system (including antennas and transmission lines) and is characterized by thermal noise (also called as white noise and Gaussian noise). External noise consists of a natural static that is primarily produced by a celestial body, such as the sun, cosmic microwave background, earth's atmosphere, electromagnetic radiation and lightning on the earth's surface, and artificial radio noise generated by numerous sources such as industries, transportation, power transmission, electrical equipment, and appliances. The artificial radio noise is generally the non-Gaussian type of noise with

Z. Zhao et al., *Radio Wave Propagation of Satellite Systems*,
https://doi.org/10.1007/978-981-97-1394-3_6

characteristics of reference impulse. The noise signals vary significantly with the different frequencies and spatiotemporal positions.

Natural and artificial radio noise signals are primarily produced from the following sources:

1. Natural noise source:

(1) Radio noise signals of atmosphere mainly arise from the electromagnetic radiation due to lightning discharge.
(2) Electromagnetic radiation from atmosphere, clouds, and rain is generally described as the "brightness temperature."
(3) "Brightness temperature" also represents the radiance of the microwave radiation from earth's surface, including ground and water surface.
(4) Noise from celestial bodies and distant galaxies, including electromagnetic radiation from the sun, moon, planets, interstellar medium, and countless galaxies in deep space.

2. Artificial noise:

(1) Electrical radiation arising from power lines mainly comprises electromagnetic radiation emitted due to gap breakdown and brush discharge, and its noise frequency spectrum is generally below 10 MHz.
(2) Electrical radiation emitted from ignition devices of internal combustion engines, mainly including the spark discharge radiation coming from the ignition system of vehicles, ships, and aircrafts, with a wide noise frequency spectrum ranging from MF to VHF. It should be noted that such noise tends to be larger than the cosmic noise in case of frequency being larger than 20 MHz.
(3) Radiation generated by electrical railways is mainly produced by the electrical spark when the pantograph mounted on these locomotives jumps off the overhead line.
(4) Leakage of industrial and medical devices.
(5) Radiation from various electrical appliances and lighting equipment used in daily life.

6.2 Solar and Cosmic Noise

The cosmic noise refers to the collective galactic noise generated by various electromagnetic radiations present in the cosmic space approaching the ground. These radiation sources consist of celestial bodies, such as the sun, moon, planets, and interstellar cloud bodies like nebulae that emit electromagnetic waves, and generate strong radiations over a wide frequency band. In general, the satellite communication system operating at 2 GHz or below should most definitely make allowances for radio noise produced by nebulae and other extraterrestrial clouds in our solar and galactic system.

6.2.1 Galactic Noise

Disregarding the shadowing effect of ionosphere, galactic noise median F_{am} (dB) can be calculated using the following formula:

$$F_{am} = 52 - 23\lg f \quad 30\,\text{MHz} \leq f \leq 100\,\text{MHz} \tag{6.1}$$

In the formula: f refers to frequency (Hz).

The cosmic microwave background (CMB) noise is about 2.73 K, but the galactic nebula appears as a narrow zone with slightly enhanced brightness, and hence only the sun and a few very strong heat exchangers should be considered in case of 2 GHz frequency and above, such as Cassiopeia A, Cygnus A and X, and Crab Nebula. The brightness temperature of extraterrestrial noise resource is as shown in Fig. 6.1, when the frequency ranges between 100 MHz and 100 GHz.

The conversion of brightness temperature of radiation from cosmic microwave background at different frequencies can be represented by the following formula (K):

$$t_b(f_i) = t_b(f_0)(f_i/f_0)^{-2.75} + 2.7 \qquad \text{k} \tag{6.2}$$

In the formula: f_0 refers to the reference frequency (MHz) and f_i refers to frequency to be converted (MHz).

The typical change of the entire frequency scope and the sky must be considered for a more precise computation. For a point resource, frequency-varying brightness temperature intensity depends on its diverse physical conditions.

6.2.2 Noise Radiation from Celestial Bodies

The sun is a strong noise source emitting radiations of different wavelengths. Its noise temperature is about 10^6 K when frequency ranges from 50 to 200 MHz; the minimum noise temperature of an inactive sun is about 10^4 K when frequency is about 10 GHz, and the noise multiplies tenfold in case of the occurrence of a solar flare.

The sun traces the ecliptic as its apparent path through the constellations. During vernal and autumnal equinox every year, the sun crosses the surface of the terrestrial equator, and GEO (geosynchronous equatorial orbit) is placed between the sun and earth, while the ground receiving antennas of the satellites are positioned opposite to the sun. At this moment, the sun turns into a huge noise source and completely overwhelms the satellites' signals leading to severe interruption of these communication systems, such as transit interruption. Transit interruption naturally happens at a fixed time every year, and hence it can be predicted accurately, to take timely countermeasures.

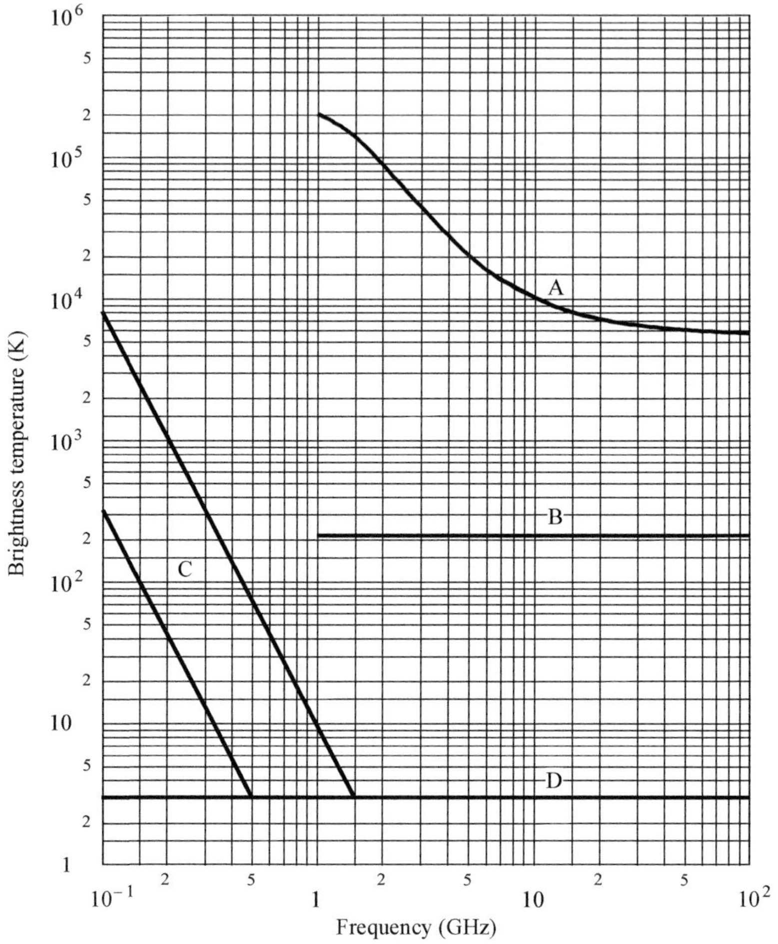

Fig. 6.1 Brightness temperature of extraterrestrial noise resources [1]; A—inactive sun; B—moon; C—range of galactic noise; D—cosmic noise

The moon intersects the ecliptic within a ±5° of declination range. When the frequency is higher than 1 GHz, the brightness temperature of moon's radiation is frequency independent. Additionally, the brightness temperature varies with time from 140 K for crescent moon to 280 K for full moon.

6.3 Noise from Earth's Surface and Atmospheric Noise

6.3.1 Uplink Noise

For uplink noise, the earth itself becomes the main noise resource. When observing from space, the average noise temperature of the earth is 254 K. When an antenna beam points to different regions of the earth, the noise temperature varies due to the existence of different mediums and terrains. On the whole, the noise temperature of earth's surface recorded by satellites is much higher than the temperature of the troposphere atmosphere, clouds, and rain.

The brightness temperature of earth's surface encompasses the radiation emitting from earth's surface as well as the reflected part of the earth's downward radiation to the atmosphere. The brightness temperature of earth's surface corresponding to a specific zenith angle is calculated using the following formula:

$$T = \varepsilon T_{\text{surf}} + \rho T_{\text{atm}} \tag{6.3}$$

In the formula: T refers to brightness temperature of earth's surface (K), ε refers to equivalent radiation coefficient of ground surface, ρ refers to equivalent reflection coefficient of ground, T_{surf} refers to the physical temperature of the earth's surface (K), and T_{atm} refers to the weighted average of brightness temperature in clear sky (K).

When frequency is 100 GHz or below, especially 10 GHz or below, the equivalent reflection coefficient of ground ρ is generally quite high, and equivalent radiation coefficient of ground surface ε is relatively low. This calculation includes all the integrations of radiation with downward angle and is delimited by the attenuation of the atmosphere.

1. Sea surface noise:

Figure 6.2a, b depicts the sea surface radiation coefficient and brightness temperature change under different cases, respectively, among which Fig. 6.2a shows radiation coefficient of smooth sea surface and vertical and horizontal polarization components of brightness temperature for two incident angles (0 and 45 m). The equivalent radiation coefficients of fresh water and sea water are identical in case of frequency being larger than 5 GHz. Figure 6.2b illustrates the relationship between nadir brightness temperature of sea surface and physical temperature of sea surface at three given frequencies (1.4, 10, and 37.5 GHz).

At the same time, the brightness temperature of sea surface increases with the increase in wind speed on the sea surface.

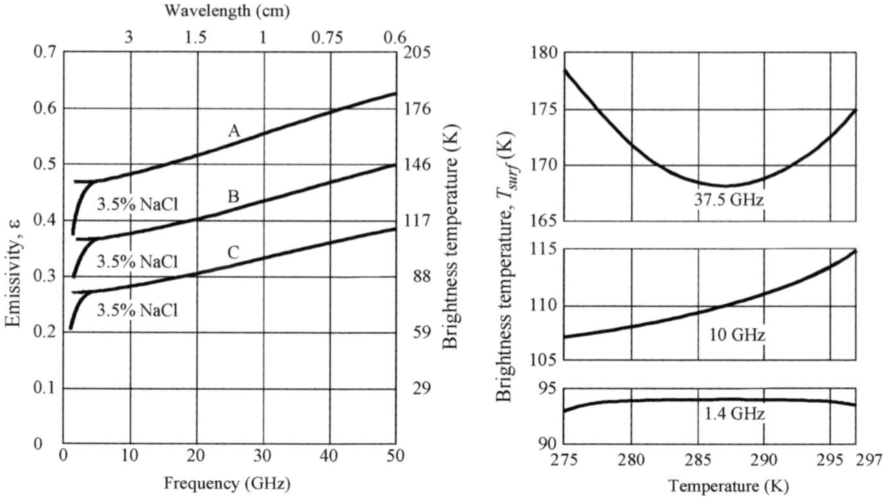

Fig. 6.2 Change of radiation coefficient and brightness temperature of sea surface (left) Radiation of smooth sea surface (% is the salt content); (right) different sea surface temperature and nadir brightness temperature at different frequencies (salt content is 3.6%). A—vertical polarization; B—0° and 45° incident angle; C—horizontal polarization

2. Land noise:

The radiation brightness temperature of land is higher than the water surface since the dielectric constant of land is low. Figure 6.3a shows the radiation brightness temperature of bare smooth land with different water content; Figure 6.3b shows the radiation brightness temperature of land with different roughness.

The figure demonstrates vertical, horizontal, and circular polarization. The radiation brightness temperature reduces with the increment in vapor density, and hence it rises accordingly with the increase in ground roughness.

For synchronous orbit satellites, wherein the beams cover the earth (the earth is within 3 dB of its main lobe), the weighted brightness temperature of earth is as shown in Fig. 6.4, in which the x-coordinate represents the longitude of satellite. The vapor density is 2.5 g/m^3, cloud coverage is 50%, lobe figure of earth coverage is provided by $G(\phi) = -3(\phi/8.715)2$ dB, and θ is standard angle with $0° \leq \varphi \leq 8.715°$.

As illustrated in Fig. 6.4, the noise temperature of the African continent near longitude 30°E is high, but the noise temperature of the region located between 180°W and 150°W is low due to the cooling influence of the Pacific Ocean. At the same time, the radiation brightness temperature increases with the increase of frequency, due to the increase in atmospheric absorption.

6.3.2 Downlink Noise

1. Atmosphere noise in clear air:

Figure 6.5 depicts the gas brightness temperature of the atmosphere as received by the ground receiver, which removes the influence of 2.73 K galactic background noise and other celestial bodies and includes the radiation brightness temperature of atmosphere at different frequencies within 1–340 GHz. Figure 6.6 shows the magnified part of 1–60 GHz in Fig. 6.5. The curve in this figure is derived using the radiation conversion formula. The input parameters of the atmosphere model are as follows: vapor density of surface is 7.5 g/m^3, surface temperature is 288 K, and vapor elevation is 2 km. The radiation brightness temperatures of 7 typical elevation angles at 0°, 5°, 10°, 20°, 30°, 60°, and 90° are provided in the figure, where the surface temperature is 15 °C and pressure is 1023 hPa.

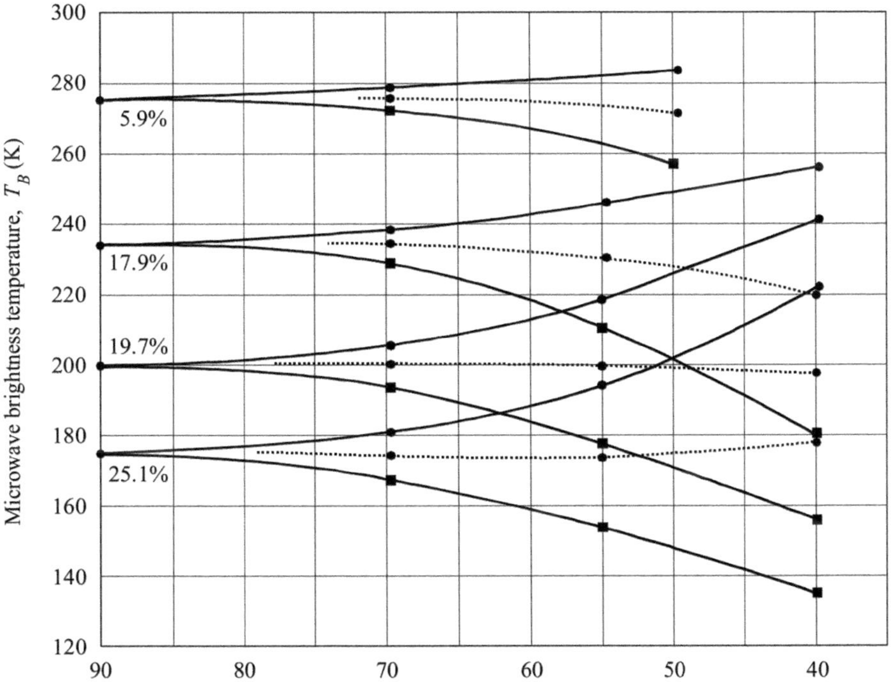

a) Moisture content from 5.9% to 25.1% for a bare, smooth field

Fig. 6.3 Relationship between 1.43 GHz land brightness temperature and elevation angle [1] **a** bare smooth land; **b** A—smooth land; B—Moderately rough land; C—rough land (deep tillage area)

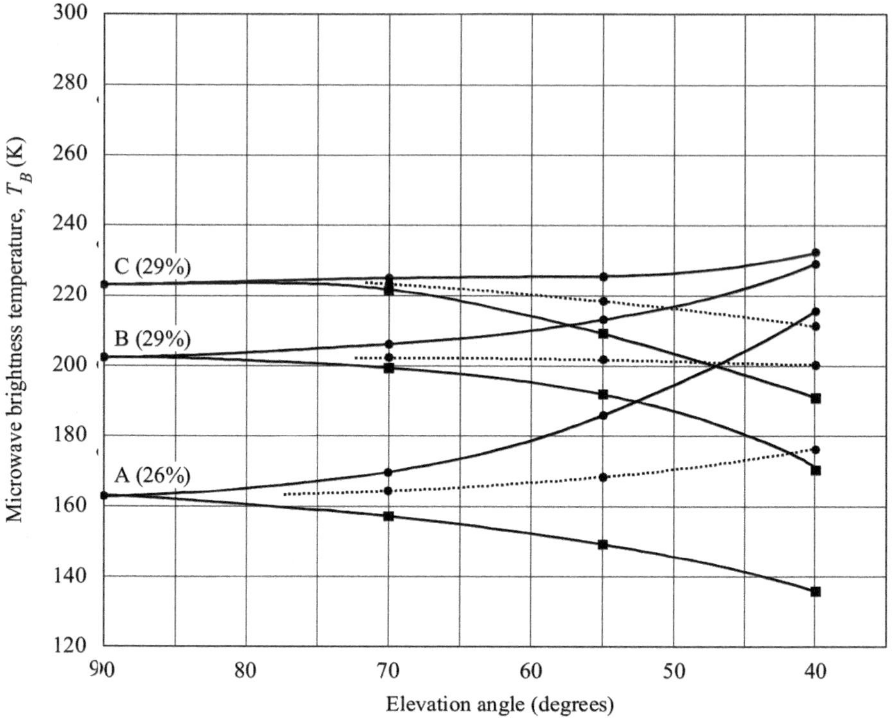

b) For (A) smooth field (B) medium roughness and (C) rough (deeply ploughed) field with percentage water content as indicated

Fig. 6.3 (continued)

The United States developed the research on radiation transfer characteristics of cloud effect, and examined the statistical characteristics of zenith brightness temperature adopting the meteorological data comprising of 15-year observation data from 15 stations. Figure 6.7 depicts the statistical distribution of radiation brightness temperature at the zenith, of two American stations (Station Yuma Arizona (1961, annual precipitation is 5.5 cm) and Station New York (1959, annual precipitation is 98.5 cm)), at 5 different frequencies (10, 18, 32, 44, and 90 GHz).

As illustrated in Fig. 6.7, the temperature of 90 GHz zenith noise may be lower than that of 44 GHz noise when the brightness temperature is very low, that is, the vapor density is very low (below 3 g/m^3); the brightness temperature of 90 GHz noise is almost identical to that of 44 GHz noise when the vapor density is 7.5 g/m^3.

2. Noise of cloud and rain:

For the ground station arranged with low-noise receiver front end, the influence of noise temperature rise on the signal-to-noise ratio may turn out to be more severe than attenuation. The window frequency band of atmospheric absorption is always

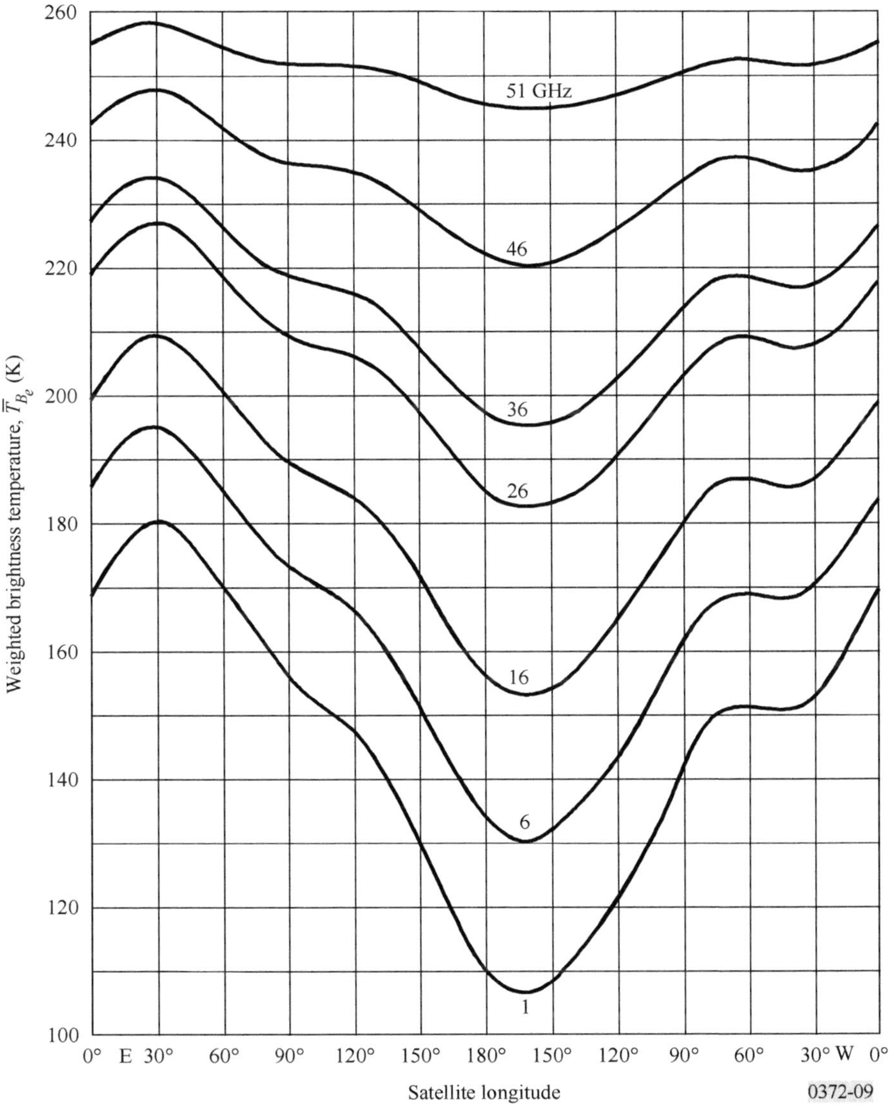

Fig. 6.4 Weighted brightness temperature of earth at 1.5 GHz synchronous orbit satellite at different longitudes [1]

employed for satellite communication. At these frequency bands, the absorption attenuation of atmosphere is minimal and the corresponding noise is small as well, since it mainly comes from clouds and rainfall. At this moment, the noise temperature of ground station antenna noise can be calculated using the following calculation:

$$T_{\text{sky}} = T_{\text{mr}}(1 - 10^{-A/10}) + 2.7 \times 10^{-A/10} \tag{6.4}$$

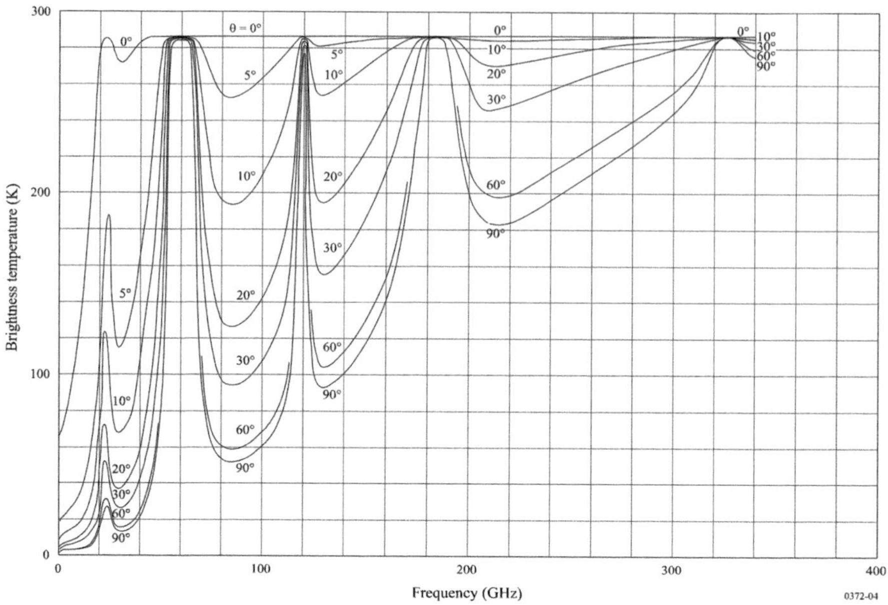

Fig. 6.5 Brightness temperature of atmospheric radiation in clear air at vapor density of 7.5 g/m^3 (1–340 GHz) [2]

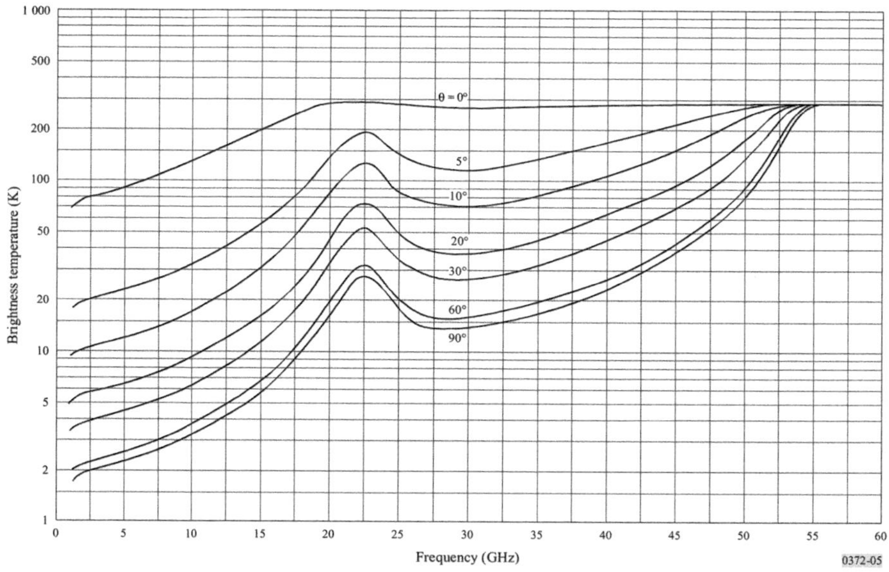

Fig. 6.6 Brightness temperature in clear air (1–60 GHz) [2]

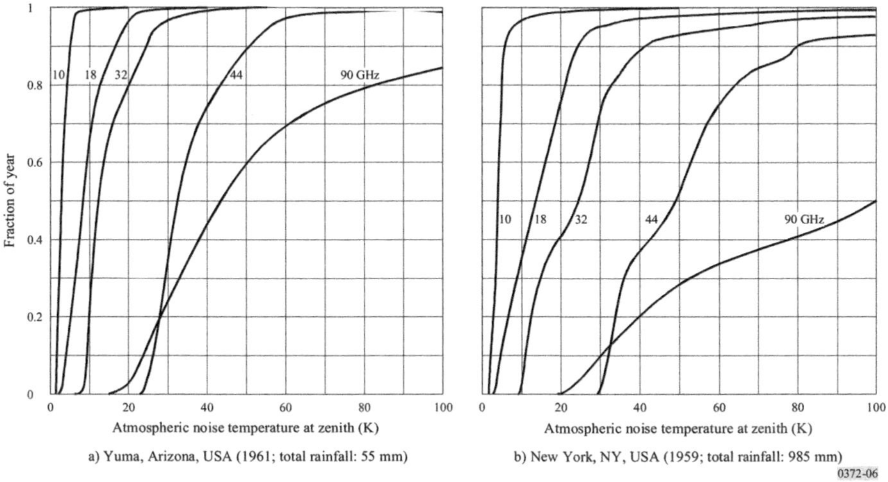

Fig. 6.7 Time probability of zenith brightness temperature is lower than x-coordinate value in a typical year [2]

In the formula: T_{sky} refers to atmospheric noise temperature relative to the antenna (K), A refers to path attenuation (dB), and T_{mr} refers to average radiation temperature of atmosphere (K).

When surface temperature T_s is known, average radiation temperature of the atmosphere can be calculated using the following formula:

$$T_{\text{mr}} = 37.34 + 0.81 \times T_S \ \text{K} \tag{6.5}$$

where T_{mr} can approximately be 275 K in case of limited local data.

6.4 Coordination Area of Earth Station

Typically, more than 1 GHz is applied for operating satellite communication network and ground microwave relay communication network. To ensure transmission quality of each network, the interference between these two communication networks must be analyzed even before the establishment of a new network. If the evaluated interference exceeds the value specified by radio regulation, the coordination must be calibrated and corresponding countermeasures must be taken to ensure the proper set-up and functioning of a new network. Otherwise, the normal functioning of the new network could suffer a failure due to the extensive interference. Therefore, evaluation and coordination of interference is undoubtedly instrumental for the establishment of a satellite communication network.

When satellite communication shares 1–40 GHz frequency band with ground microwave relay communication network, the transiting signal of the relay could interfere with the receiving signal of the ground station, and the transiting signal of the ground station may interfere with the receiving signal of the ground relay. To retain normal communication, a certain distance must be maintained between the earth station and the ground relay. The interference can be overlooked when exceeding such distance, and this distance is called as coordination distance. The area formed by connecting the endpoints of coordination distance at each azimuth of the earth station is regarded as the coordination area. The coordination distance should be defined before determining the coordination area of an earth station. Therefore, various factors must be considered, such as transmission loss of interference signal, atmospheric influence, terrain near station, direction of antenna, and time probability of the occurrence of tolerable interference signal.

6.4.1 Coordination Distance

There exist two transmission models that should be considered for computing the interference between earth station of satellite communication network and earth station of ground microwave relay communication network, namely, clear-air propagation model and rainfall scattering model. The method for calculating the coordination distance is provided in subsequent paragraphs.

1. Overview:

When calculating the coordination distance using clear-air propagation model, different calculation methods should be used for the three frequency bands mentioned below:

(1) The empirical propagation method that is suitable with respect to the measured data should be used for 100–790 MHz VHF/UHF frequency bands.
(2) Effect of troposphere scattering, atmospheric duct, and stratification reflection/refraction should be considered for 790 MHz to 60 GHz.
(3) Vapor absorption effect pertaining to free-space propagation loss, and conservation and signal enhancement with small time probability must be considered for 60–105 GHz.

When determining the coordination distance while considering rainfall scattering, the isotropic scattering produced in public scatter by the main beam of each interference station due to rainfall should be considered. When performing frequency coordination, the interference of rainfall to frequencies beyond 1–40.5 GHz can be ignored. This is because the level of rainfall scattering signal below 1 GHz is very low, and although significant scattering occurs when the frequency is above 40.5 GHz, the scattering signal undergoes large rain attenuation in the path from scatterer to ground station.

2. Determination of limit of coordination distance:

The maximum and minimum coordination distances must be first determined in calculating the coordination distance. The distance should increase from minimum and not exceed the maximum when calculating the coordination distance of clear-air effects. While considering rainfall scattering effect, the distance must decrease from maximum and not be less than minimum.

(1) Minimum coordination distance:

The minimum coordination distance of rainfall scattering effect is about 55 km. The minimum coordination distance of clear-air effects $d_{\min}$ (km) is calculated using the following formula:

$$d_{\min}(f) = \begin{cases} d'_{\min}(f) & 0 < f < 40\,\text{GHz} \\ \frac{(54-f)d'_{\min}(40)+10(f-40)}{14} & 40\,\text{GHz} \le f < 54\,\text{GHz} \\ 10 & 54\,\text{GHz} \le f < 66\,\text{GHz} \\ \frac{10(75-f)+45(f-66)}{9} & 66\,\text{GHz} \le f < 75\,\text{GHz} \\ 45 & 75\,\text{GHz} \le f < 90\,\text{GHz} \\ 45 - \frac{(f-90)}{1.5} & 90\,\text{GHz} \le f \le 105\,\text{GHz} \end{cases} \tag{6.6}$$

$$d'_{\min}(f) = 100 + \frac{(\beta_p - f)}{2} \tag{6.7}$$

In the formula: f is frequency (GHz), $d'_{\min}$ (40) refers to minimum coordination distance (km) considering propagation in clear air when f is 40 GHz, and frequency f for above formula is limited within 40 GHz.

β_p refers to parameter of radio meteorology, calculated using the following formula:

$$\beta_p = \begin{cases} 10^{1.67-0.015\zeta_r} & \zeta_r \le 70^\circ \\ 4.17 & \zeta_r > 70^\circ \end{cases} \tag{6.8}$$

In the formula: ζ_r is determined by latitude of the earth station (°):

$$\zeta_r = \begin{cases} |\zeta| - 1.8 & |\zeta| > 1.8^\circ \\ 0 & |\zeta| \le 1.8^\circ \end{cases}$$

(2) Maximum coordination distance:

In an iterative calculation of coordination distance considering clear-air effects and rainfall scattering effect, it is necessary to fix the maximum coordination distance $d_{\max1}$ (km) and $d_{\max2}$ (km), respectively. $d_{\max1}$ is calculated using the following formula:

$$d_{\max1} = \begin{cases} 1200 & f \le 60\,\text{GHz} \\ 80-10\log\left(\frac{p_1}{50}\right) & f > 60\,\text{GHz} \end{cases} \tag{6.9}$$

Table 6.1 Maximum coordination distance under rainfall scattering effect

Latitude/°	0–30	30–40	40–50	50–60	360
Distance/km	350	360	340	310	280

$d_{\max 2}$ is calculated using Table 6.1.

6.4.2 *Coordination Distance Considering Propagation in Clear Air*

1. Classification of radio meteorological zones:

By and large, the globe is divided into four radio meteorological zones for analyzing the coordination distance:

(1) Zone B: Cold seas, oceans, and large inland waters at latitudes greater than 30°, excluding the Mediterranean Sea and Black Sea.
(2) Zone C: Warm seas, oceans, and large inland waters at latitudes lower than 30°, including Mediterranean Sea and Black Sea.
(3) Zone A1: Coastal lands and areas, that is, areas adjacent to Zone B or C up to areas 100 m above sea or water level, with a maximum distance of 50 km from the nearest Zone B or C; the approximate value can be applied if there is no accurate data available about the 100 m elevation contour.
(4) Zone A2: All land, excluding coastal land and areas as defined in Zone A1 above.

The following approximations are required for large inland waters, inland lakes, or high-wetland areas: Large inland waters refer to waters covering an area of at least 7,800 km^2 in Zone B or C, excluding river areas. Islands in such waters are classified for calculation if 90% of island area is not more than 100 m above the mean water level. For calculation of waters, the islands that do not satisfy these rules are classified as lands. When a wetland area larger than 7,800 km^2 contains many small lakes or river networks, such an area is classified as coastal area A1, if more than 50% of its area is water and more than 90% of its land is not more than 100 m above the mean water level.

2. Worst-month coordination distance:

The coordination distance is usually determined by the interference level with $p_1\%$ time probability not exceeding annual average. P_1 can be calculated using the following formula when coordination distance is required not to exceed the worst-month $p_{w1}\%$ time probability:

$$p_1 = 10^{\frac{\log(p_{w1}) + \log(G_L) - 0.444}{0.816}} \tag{6.10}$$

Table 6.2 Determination of radio horizontal distance

Horizontal distance range	Horizontal distance
$d_h < 0.5$ km	0.5 km
0.5 km $\leq d_h \leq$ 5.0 km	Horizontal distance
$d_h > 5.0$ km	5.0 km

In which,

$$G_L = \begin{cases} \sqrt{1.1 + |\cos 2\zeta_r|^{0.7}} & \zeta_r \leq 45^\circ \\ \sqrt{1.1 - |\cos 2\zeta_r|^{0.7}} & \zeta_r > 45^\circ \end{cases} \tag{6.11}$$

and p_1 should satisfy the condition of $12p_1 \geq p_{w1}$.

3. Calculation of coordination distance considering propagation in clear air:

Bearing in mind the clear-air effects, parameter A_h is used to measure the shadowing of earth station due to its surrounding terrain. The additional loss due to the shadowing is calculated along each direction. Horizontal distance of radio d_h is defined in accordance with Table 6.2. Calculate the horizontal angle θ_h (°) between horizontal plane and ray across the shadowing, taking earth station as its vertex. θ_h is a positive value when shadowing is above water level. The horizontal angle should be determined for all the directions surrounding the earth station. In fact, 5° of increment is deemed sufficient for the calculation. However, it is required to determine whether there is a minimum horizontal angle in 5° of increment for each calculation.

Calculating the horizontal distance correction A_d (dB) along each azimuth surrounding the earth station:

$$A_d = 15\left[1 - \exp\left(\frac{0.5 - d_h}{5}\right)\right]\left[1 - \exp\left(-\theta_h f^{1/3}\right)\right] \tag{6.12}$$

Transmission loss A_h (dB) caused by terrain shadowing at each azimuth of the earth station:

$$A_h = \begin{cases} -1.5\left[(f+1)^{1/2} - 0.0001f - 1.0487\right] & \theta_h < -0.5^\circ \\ 3\left[(f+1)^{1/2} - 0.0001f - 1.0487\right]\theta_h & -0.5^\circ < \theta_h < 0^\circ \\ 20\lg\left(1 + 4.5\theta_h f^{1/2}\right) + \theta_h f^{1/3} + A_d & \theta_h \geq 0^\circ \end{cases} \tag{6.13}$$

and A_h value must satisfy the following conditions:

$$-10 \leq A_h \leq (30 + \theta_h) \tag{6.14}$$

The determination methods for coordination distance of satellite communication system at different frequency bands are described in the subsequent paragraphs.

(1) 100–790 MHz:

Calculate and iterate basic transmission loss using different path types starting from minimum distance. Distance at the beginning of each iterative calculation is

$$d_i = d_{\min} + i \cdot s\,(i = 0, 1, 2, \ldots) \tag{6.15}$$

The transmission losses of various path types are calculated using the following formula:

$$\begin{aligned} L_{bl}(p_1) &= 142.8 + 20\lg f + 10\lg p_1 + 0.1d_i \quad \text{(All paths are in Zone A1 or A2)} \\ L_{bs}(p_1) &= 49.91\lg\left(d_i + 1840f^{1.76}\right) + 1.195f^{0.393}(\lg p_1)^{1.38}d_i^{0.597} \\ &\quad + (0.01d_i - 70)(f - 0.1581) + \left(0.02 - 2\times10^{-5}p_1^2\right)d_i \\ &\quad + 9.72\times10^{-9}d_i^2p_1^2 + 20.2 \quad \text{(All paths are in Zone B)} \end{aligned}$$

$$\begin{aligned} L_{bs}(p_1) &= 49.343\lg\left(d_i + 1840f^{1.58}\right) + 1.266(\lg p_1)^{(0.468+2.598f)}d_i^{0.453} + (0.037d_i - 70)(f - 0.1581) \\ &\quad + 1.95\times10^{-10}d_i^2p_1^3 + 20.2 \quad \text{(All paths are in Zone C)} \end{aligned}$$

The basic transmission loss under current distance is calculated using the following formula:

$$L_2(p_1) = L_{bs}(p_1) + \left[1 - \exp\left(-5.5\left(\frac{d_{tm}}{d_i}\right)^{1.1}\right)\right](L_{bl}(p_1) - L_{bs}(p_1)) \tag{6.16}$$

In the formula: d_{tm} (km) refers to the distance of largest continuous land (inland coastal land) in Zone A1 and A2 under current distance.

The iteration is required to satisfy the following relational expression:

$$L_2(p_1) \geq L_1(p_1) \tag{6.17}$$

or

$$d_i \geq d_{\max 1}$$

If the recommended distance increment s is 1 km, then:

$$L_1(p_1) = L_b(p_1) - A_h \tag{6.18}$$

In the formula: L_1 (p_1) (dB) refers to minimum basic transmission loss required by $p_1\%$ time probability exceeded.

This calculation method is applicable for $p_1\%$ time probability within a range of 1–5%. So far, this iteration is complete. Then, comparing the relationship between $L_2(p_1)$ and $L_1(p_1)$ and between d_i and $d_{\max 1}$, to decide whether the iteration should be processed or terminated.

(2) 790–60,000 MHz:

The troposphere scattering, atmospheric duct, and stratification reflection/refraction should be considered for this frequency band. The propagation model here can be applied into the annual time probability within 0.001–50%. Evaluate using the said iteration method. Iterating the distance d_i ($i = 0, 1, 2 \ldots$) by adopting a uniform step size starting from minimum coordination distance $d_{\min}$.

In which:

$$d_i = d_{\min} + i \cdot s \tag{6.19}$$

At distance d_i at each azimuth angle of the earth station, the vapor density ρ_i (g/m^3) with 50% time probability exceeded can be obtained from ITU-R.836 Recommendation. Then, calculating the distance-dependence attenuation caused by vapor absorption:

$$A_g = (\gamma_o + \gamma_d)d_i + \sum_{n=0}^{i} \gamma_w(\rho_n) \cdot s \tag{6.20}$$

In the formula: γ_d refers to attenuation rate of atmospheric duct (dB/km), being defined as

$$\gamma_d = 0.05 f^{1/3} \tag{6.21}$$

where γ_o refers to attenuation rate of dry air (dB/km), being defined as

$$\gamma_o = \begin{cases} \left[7.19 \times 10^{-3} + \frac{6.09}{f^2 + 0.227} + \frac{4.81}{(f-57)^2 + 1.50} f^2 \times 10^{-3}\right] & f \leq 56.77 \\ 10 & f > 56.77 \end{cases} \tag{6.22}$$

where γ_w refers to attenuation rate of vapor (dB/km), being defined as

$$\gamma_w(\rho) = \left(0.050 + 0.0021\rho + \frac{3.6}{(f - 22.2)^2 + 8.5}\right) f^2 \rho \times 10^{-4} \tag{6.23}$$

The vapor density used for vapor attenuation rate in the troposphere scatter propagation is 3.0 g/m^3, then

$$\gamma_{wt} = \gamma_w(3.0) \tag{6.24}$$

Get vapor density ρ_0 at the earth station and vapor density $\rho_{d\min}$ at distance $d_{\min}$ at each azimuth angle from ITU-R.836 Proposal. Calculate region-dependent parameters:

$$\tau = 1 - \exp\left[-\left(4.12 \times 10^{-4} d_{lm}^{2.41}\right)\right] \tag{6.25}$$

$$\mu_1 = \left[10^{\frac{-d_{tm}}{16-6.6\tau}} + \left[10^{-(0.496+0.354\tau)}\right]^5\right]^{0.2} \tag{6.26}$$

$\mu_1 \le 1$.

$$\sigma = -0.6 - \varepsilon_L \times 10^{-9} d_i^{3.1} \tau \tag{6.27}$$

$\sigma \ge -3.4$.

$$\mu_2 = \left(2.48 \times 10^{-4} d_i^2\right)^{\sigma} \tag{6.28}$$

$\mu_2 \le 1$.

$$\mu_4 = \begin{cases} 10^{(-0.935+0.0176\zeta_r)\log \mu_1} & \zeta_r \le 70^\circ \\ 10^{0.3\log \mu_1} & \zeta_r > 70 \end{cases} \tag{6.29}$$

Calculate the occurrence rate of atmospheric duct β and relevant parameters Γ, as both are used for calculation of time dependence of basic transmission attenuation.

$$\beta = \beta_p \cdot \mu_1 \cdot \mu_2 \cdot \mu_4 \tag{6.30}$$

$$\Gamma = \frac{1.076}{(2.0058 - \log \beta)^{1.012}} \exp\left[-\left(9.51 - 4.8\log \beta + 0.198(\log \beta)^2\right) \times 10^{-6} d_i^{1.13}\right] \tag{6.31}$$

Calculate distance-dependent attenuation in atmospheric duct:

$$L_5(p_1) = A_g + \left(1.2 + 3.7 \times 10^{-3} d_i\right)\log\left(\frac{p_1}{\beta}\right) + 12\left(\frac{p_1}{\beta}\right)^{\Gamma} \tag{6.32}$$

Calculate distance-dependent attenuation in tropospheric scattering:

$$L_6(p_1) = 20\log d_i + 5.73 \times 10^{-4}(112 - 15\cos(2\zeta))d_i + (\gamma_o + \gamma_{wt})d_i \tag{6.33}$$

(1) Considering the occurrence of atmospheric duct:

Calculate attenuation correction (dB) of atmospheric duct for the sea surface:

$$A_c = \frac{-6}{(1 + d_c)} \tag{6.34}$$

In the formula: d_c (km) refers to distance from land-based earth station to the coast in considered direction, and d_c is zero in other environments.

Calculate non-distance-dependent attenuation (dB):

$$A_1 = 122.43 + 16.5 \log f + A_h + A_c + A_w \tag{6.35}$$

Calculate minimum distance-dependent attenuation (dB):

$$L_3(p_1) = L_b(p_1) - A_1 \tag{6.36}$$

Set a factor to control other attenuations related to the additional links and the terrain height:

$$\varepsilon_L = 8.5 \tag{6.37}$$

(2) Considering the tropospheric scattering:

Calculate frequency-dependent loss (dB):

$$L_f = 25 \log f - 2.5\left[\log\left(\frac{f}{2}\right)\right]^2 \tag{6.38}$$

Calculate non-frequency-dependent attenuation (dB):

$$A_2 = 187.36 + 10\theta_h + L_f - 0.15N_0 - 10.1\left(-\log\left(\frac{p_1}{50}\right)\right)^{0.7} \tag{6.39}$$

In the formula: θ_h refers to horizontal elevation angle of earth station (°) and N_0 refers to refraction rate of sea surface at the center of the path.

Calculate minimum distance-dependent attenuation (dB):

$$L_4(p_1) = L_b(p_1) - A_2 \tag{6.40}$$

Calculate using formula (6.22) to formula (6.40) in iteration process, and then iterate the distance d_i ($i = 0, 1, 2 \ldots$) by adopting proper increment step as per formula (6.19) to formula (6.33) starting from minimum coordination distance $d_{\min}$. d_i is regarded as the current distance for each iteration. Continue this process until both the following expressions are established:

$$L_5(p_1) \geq L_3(p_1) \tag{6.41}$$

$$L_6(p_1) \geq L_4(p_1) \tag{6.42}$$

Or

$$d_i \geq d_{\max 1}$$

Subsequently, get coordination distance d_1 from the last iteration. This calculation method is applicable for annual time probability range within 0.001–50%. The recommended distance increment s is 1 km.

(3) 60–105 GHz:

Within millimeter-wave frequency range from 60 to 105 GHz, the free-space attenuation and vapor absorption must be considered and signal enhancement with small time probability should be added as well.

Calculate the attenuation rate of dry air (dB/km) within 60–105 GHz:

$$\gamma_{om} = \begin{cases} \left[2 \times 10^{-4}\left(1 - 1.2 \times 10^{-5} f^{1.5}\right) + \frac{4}{(f-63)^2+0.936} \right. \\ \left. + \frac{0.28}{(f-118.75)^2+1.771}\right] f^2 6.24 \times 10^{-4} & f > 63.2\,\text{GHz} \\ 10\,\text{dB/km} & f \leq 63.26\,\text{GHz} \end{cases} \tag{6.43}$$

Calculate vapor absorption at vapor density of 3 g/m^3 using the following formula:

$$\gamma_{wm} = \left(0.039 + 7.7 \times 10^{-4} f^{0.5}\right) f^2 2.369 \times 10^{-4} \tag{6.44}$$

Calculate attenuation of atmosphere absorption:

$$\gamma_{gm} = \gamma_{om} + \gamma_{wm} \tag{6.45}$$

Calculate A_h using formula (6.13), and then measure non-distance-dependent basic transmission loss:

$$L_7 = 92.5 + 20 \log f + A_h \tag{6.46}$$

Calculate minimum distance-dependent loss (dB):

$$L_8(p_1) = L_b(p_1) - L_7 \tag{6.47}$$

Calculate the iteration process.

Each iteration distance is ($i = 0, 1, 2,$)

$$d_i = d_{\min} + i \cdot s \tag{6.48}$$

Calculate distance-dependent loss of the current distance:

$$L_9(p_1) = \gamma_{gm} d_i + 20 \log d_i + 2.6\left[1 - \exp\left(\frac{-d_i}{10}\right)\right] \log\left(\frac{p_1}{50}\right) \tag{6.49}$$

It is required to satisfy the following condition:

$$L_9(p_1) \geq L_8(p_1) \tag{6.50}$$

Or

$$d_i \geq d_{\max 1} \tag{6.51}$$

Therefore, the distance obtained by the last "i" is the required coordination distance. This calculation method is suitable for $p_1\%$ time probability in the range of 0.001–50%.

6.4.3 *Coordination Distance Considering Hydrometeor Scattering*

The determination of coordination distance considering hydrometeor scattering is analyzed based on geometric relationship of paths different from the circular propagation mechanism of the earth. Hydrometeors cause scattering of isotropic energies, thus causing interference with large scattering angle and cause the beam to intersect with the large circular path. The division method mentioned above, stating that the earth's surface is divided into inland, coastal area, and sea surface, no longer applies to the propagation calculation.

1. Calculation of coordination distance considering hydrometeor scattering:

The determination of coordination distance considering hydrometeor scattering is based on the bistatic radar equation and adopts the "narrow beam" approximation of earth station's antenna, as the antenna gain can cancel the diffuse attenuation from the scatterer to the antenna. Therefore, this interference coordination distance is essentially decided by the path length from ground station to the scatterer, called as the rain belt. This method is applicable to 1–40.5 GHz. The interference of hydrometeor scattering can be disregarded in case this frequency range is exceeded. Determine its minimum coordination distance using formula (6.6). The geometric relation of hydrometeor scattering effect is shown in Fig. 6.8.

The transmission loss $Lr(p_2)$ is calculated using the following formula:

$$L_r(p_2) = 173 + 20\lg r_i - 20\lg f - 14\lg R(p_2) - 10\lg C + 10\lg S - G_T + A_g \tag{6.52}$$

In the formula: C refers to total effective scattering transmission function, 10lg S refers to Rayleigh scattering bias, which is only applicable for hydrometeor scattering below the edge of the rain top, A_g refers to attenuation due to atmosphere, and G_T refers to antenna gain of ground station. Calculate using the following formula:

$$C = \Gamma_b C_b + \Gamma_a C_a \tag{6.53}$$

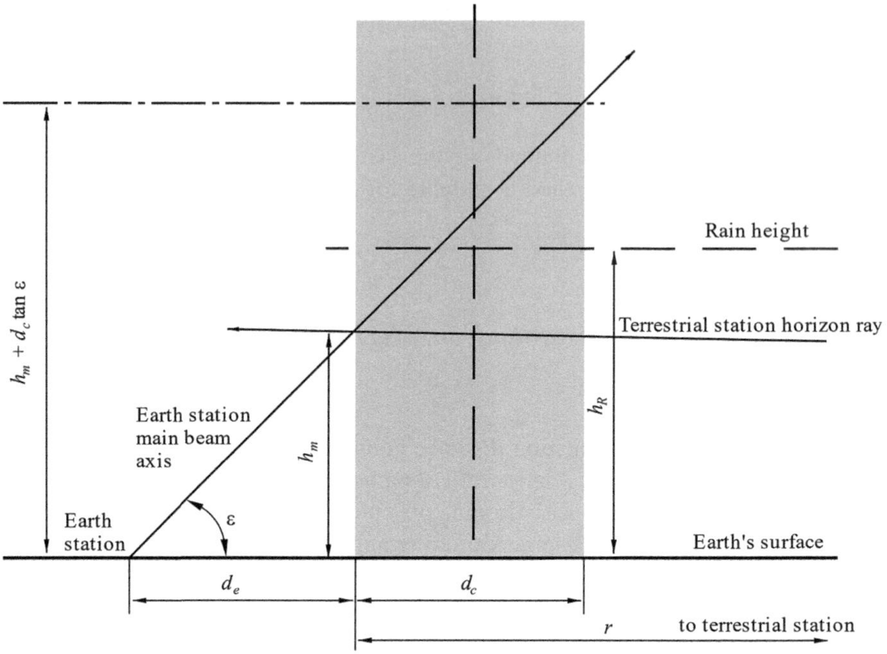

Fig. 6.8 Hydrometeor scattering effect [3]

$$10\lg S = \begin{cases} 0.005(f - 10)^{1.7} R(p_2)^{0.4} & 10\,\text{GHz} \le f \le 40\,\text{GHz} \\ 0 & f < 10\,\text{GHz Or } C_b = 0 \end{cases} \tag{6.54}$$

$$A_g = \gamma_o(d_{to} + d_{ro}) + \gamma_{wv}(d_{tv} + d_{rv}) \tag{6.55}$$

In the formula: Γ_b refers to path attenuation due to hydrometeor scattering, Γ_a refers to path attenuation due to thaw layer and ice scattering, C_b refers to effective scattering transmission function of hydrometeor scattering, and C_a refers to transmission function of thaw layer and ice scattering. Calculate using the following formula:

$$\Gamma_b = \exp\left[-0.23\left(\frac{\Gamma_1}{\cos\varepsilon} + \Gamma_2\right)\right] \tag{6.56}$$

$$\Gamma_a = \exp\left[-0.23\left(\frac{\Gamma_1}{\cos\varepsilon} + \gamma_R \frac{h_c - h_m}{\sin\varepsilon}\right)\right] \tag{6.57}$$

$$C_b = \frac{4.34}{\gamma_R(1 + \cos\varepsilon)}\left[1 - \exp\left\{-0.23\gamma_R(h_c - h_m)\frac{\sin\varepsilon}{1 - \cos\varepsilon}\right\}\right] \tag{6.58}$$

$$C_a = \frac{0.67}{\sin\varepsilon_s}\left[\exp\{-1.5(h_c - h_R)\} - \exp\{-1.5(h_m - h_R + d_c\tan\varepsilon)\}\right] \tag{6.59}$$

In formula (6.56) to formula (6.59): Γ_1 refers to attenuation from intersection of beam to earth station (dB), Γ_2 refers to attenuation from current distance point to intersection of beam (dB), ε refers to elevation angle of earth station's antenna (°), h_c refers to rain belt and region-related parameter at beam intersection, h_m refers to height of the intersection above the ground surface between antenna beam of ground station and antenna beam of earth station (km), and γ_R refers to rain attenuation rate (dB/km).

Attenuation from beam intersection to earth station Γ_1 is calculated using the formula:

$$\Gamma_1 = \begin{cases} \gamma_R r_m\left[1 - \exp\left(-\frac{d_e}{r_m}\right)\right] & h_m \le h_R \\ \gamma_R r_m\left[\exp\left(-\frac{(h_m - h_R)\cot\varepsilon}{r_m}\right) - \exp\left(-\frac{d_e}{r_m}\right)\right] & h_m > h_R \end{cases} \tag{6.60}$$

In the formula: d_e refers to horizontal distance from earth station to edge of rain belt and r_m refers to scaling distance of attenuation outside the common scatterer volume (km). Calculate using the following formula:

$$d_e = r_E \arcsin\left(\frac{r_r}{h_m + r_E}\cos\varepsilon\right) \tag{6.61}$$

$$r_m = 600R(p_2)^{-0.5} \times 10^{-\eta} \tag{6.62}$$

In the formula: $R(p_2)$ refers to rainfall rate at the earth station with $p_2\%$ time probability exceeded (mm/h), which can be determined using ITU-RP.837 Recommendation, when the local data is available; r_r refers to path length from beam intersection to earth station (km); and η refers to middle parameter.

Calculate using the following formula:

$$r_r = \sqrt{r_E^2 \sin^2\varepsilon + h_m^2 + 2h_m r_E} - r_E \sin\varepsilon \tag{6.63}$$

$$h_m = r_E\left(\frac{1}{\cos\delta} - 1\right) \tag{6.64}$$

$$\eta = (R(p_2) + 1)^{0.19} \tag{6.65}$$

In the formula: r_E refers to effective radius of earth (km) (8,500 km) and δ refers to angle (°) between rain belt and the earth surface point above the current distance r_i, which can be defined as

$$\Gamma_2 = \gamma_R r_m\left[1 - \exp\left(-\frac{r_t}{r_m}\right)\right] \tag{6.66}$$

The attenuation from current distance point to Γ_2 beam intersection is calculated using the following formula:

$$r_t = h_m\sqrt{1 + 2\frac{r_E}{h_m}} \tag{6.67}$$

In the formula: r_t refers to path length (km) from ground station to beam intersection, which can be defined as

$$\delta = \frac{r_i}{r_E} \text{ rad} \tag{6.68}$$

Rain belt and area-related parameter at beam interaction is given as

$$h_c = \begin{cases} h_m & h_R \le h_m \\ h_R & h_m < h_R < h_m + d_c \tan\varepsilon \\ h_m + d_c \tan\varepsilon & h_R \ge h_m + d_c \tan\varepsilon \end{cases} \tag{6.69}$$

In the formula: d_c refers to diameter of rain belt (km).

$$d_c = 3.3R(p_2)^{-0.08} \tag{6.70}$$

In the formula: γ_o refers to attenuation rate of dry air (dB/km) and γ_w refers to attenuation rate of vapor (dB/km). Calculate using the following formula:

$$\gamma_o = \left[7.19 \times 10^{-3} + \frac{6.09}{f^2 + 0.227} + \frac{4.81}{(f - 57)^2 + 1.5}\right] f^2 \times 10^{-3} \tag{6.71}$$

$$\gamma_{wv} = \left[0.050 + 0.0021\rho + \frac{3.6}{(f - 22.2)^2 + 8.5}\right] f^2 \rho \times 10^{-4} \tag{6.72}$$

In the formula: ρ refers to the vapor density of ground with 50% time probability exceeded (g/m^3), determined in accordance with the ITU-RP.836 Recommendation. For simplification, we can assume that the vapor density is a constant in the path when determining the contour line of coordination distance under hydrometeor scattering.

In the formula: d_{to} and d_{tv} refer to oxygen equivalent path and vapor equivalent path (km) and d_{ro} and d_{rv} refer to equivalent path from earth station to rain belt for characterizing oxygen and vapor (km), respectively. Calculate using the following formula:

$$d_{to} = \begin{cases} 0.9r_t & r_t < 270 \text{ km} \\ 243 + 0.4(r_t - 270) & r_t \ge 270 \text{ km} \end{cases} \tag{6.73}$$

$$d_{tv} = \begin{cases} 0.85r_t & r_t < 220 \text{ km} \\ 187 + 0.4(r_t - 220) & r_t \ge 220 \text{ km} \end{cases} \tag{6.74}$$

$$d_{ro} = 0.8r_r \tag{6.75}$$

$$d_{rv} = 0.5r_r \tag{6.76}$$

In the above calculation, the distance between earth station and rain belt along the horizontal direction of earth station is usually not more than 30 km, even if the satellite-earth communication elevation angle is as low as 10°; hence, no significant difference in rainfall rate and rain top height between these two points can be observed.

Through iterative calculation, get maximum distance $d_{\max2;}$ as per Table 6.3, r_i ($i = 0, 1, 2 \ldots$) refers to distance between rain belt and ground station (km):

$$r_i = d_{\max2} - i \cdot s \tag{6.77}$$

It needs to satisfy the given conditions using the following formula:

$$L_r(p_2) < L_b(p_2) \tag{6.78}$$

Or

$$r_i < d_{\min} \tag{6.79}$$

Here, the previous iteration value of r_i provides the calculation distance of rain belt scattering:

$$d_r = d_{\max2} - (i - 1) \cdot s \tag{6.80}$$

If $d_r < d_{\min}$ is obtained by iteration calculation, then $d_r = d_{\min}$ and iteration calculation is complete. For hydrometeor scattering effect, $d_{\min} = 55$ km. Assuming the scattering due to rain belt above each azimuth is isotropic, the coordination contour line is a circle of radius d_r and distance from the center to the earth station is d_e, as shown in Fig. 6.9.

2. Determination of auxiliary coordination line of hydrometeor scattering effect:

The auxiliary contour line allows the azimuth deviation of antenna beam of the ground station for position of the earth station considering their coordination. Figure 6.10 depicts the hydrometeor scattering zones projected on the horizontal plane. The earth station and ground station in the figure are located at points A and B, respectively. Position B of the ground station is in the direction determined by angle ω from point C on the main beam axis of earth station. Point C is the center of main contour line or supplementary contour line of hydrometeor scattering effect and is also the center of auxiliary contour line. The shaded area in Fig. 6.10 refers to the critical area between rainfall height and earth station along the main beam axis of the earth station. A

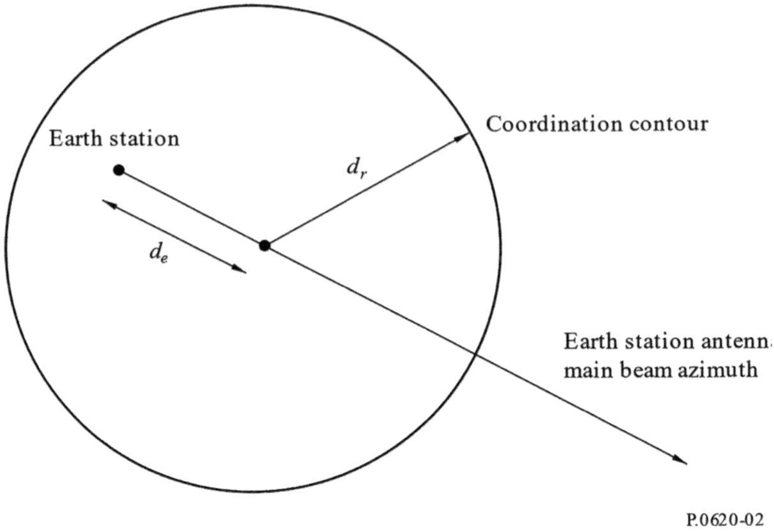

Fig. 6.9 Position of coordination contour line [3]

common body forms between the beam of earth station in this critical area and the beam of any ground station in the main contour line or supplementary contour line of the hydrometeor scattering effect. The length of this critical area is b and maximum horizontal length is up to point M. The intersection of this critical area and main beam of the ground station produces a huge rainfall scattering interference through the coupling taking place between the major lobes.

For the point specified in the main contour line or supplementary contour line of hydrometeor scattering effect, its angle to critical area is critical angle ψ. The protection angle v in Fig. 6.10 refers to the angle of main beam deviating from the critical area. The avoidance angle between main beam axis of ground station and earth station is ϕ. Angle ϕ is the sum of angle ψ and angle v, whose size in degrees has a fixed value for specific auxiliary contour line. Each auxiliary contour line is generated by changing ω from point C to the auxiliary contour and the derived distance r_b. ψ and v vary with increase of ω from 0° to 360°, but their sum does not change.

This method is based on the main contour line given by formula (6.80). Iteratively decrease the distance r_b between ground station and earth station from distance d_r, until the minimum distance r_b satisfying the required minimum loss is obtained, or the minimum coordination distance is reached. Determine critical angle ψ for each r_b value, and then calculate the protection angle v. Put the antenna gain corresponding to protection angle v and current distance r_b into formula (6.52) to get the path attenuation of hydrometeor scattering effect. Repeat the above process for each angle ω, and generate a closed auxiliary contour line for a given beam avoidance angle φ. Auxiliary coordination line of hydrometeor scattering effect is determined as per the following procedures:

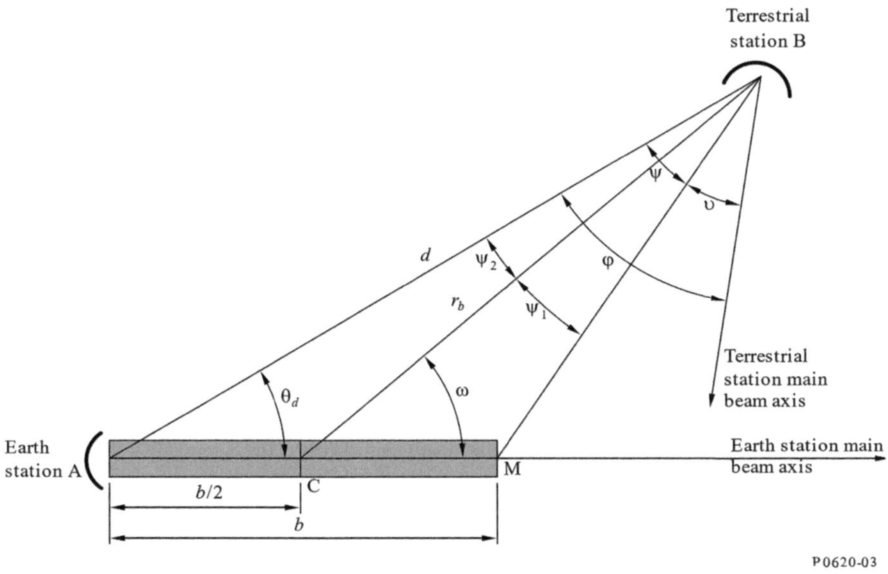

Fig. 6.10 Geometric relation of propagation in the horizontal plane [3]

① Set r_b value for radius d_r of main contour line or supplementary contour line of hydrometeor scattering effect in accordance with the calculation method described in formula (6.80).

② Calculate critical angle ψ:

$$\psi_1 = \arctan\left(\frac{b \sin \omega}{2r_b - b \cos \omega}\right) \tag{6.81}$$

$$\psi_2 = \arctan\left(\frac{b \sin \omega}{2r_b + b \cos \omega}\right) \tag{6.82}$$

$$\psi = \psi_1 + \psi_2 \tag{6.83}$$

③ If $\psi > \varphi$, the current angle ω is consistent with the auxiliary contour line and main contour line of hydrometeor scattering effect. Move to procedure step (10) after calculation of ω. Otherwise, move to procedure steps (4) to (9), until one of the final conditions described in procedure steps (6) and (3) is satisfied.

④ r_b declines by 0.2 km.

⑤ Use formula (6.81) to (6.83) to re-calculate the critical angle ψ.

⑥ If 0.5b sin ω/sin $\psi_2 < d_{\min}$, then the contour line of hydrometeor scattering effect is consistent with the minimum coordination distance, then move to procedure step (8) after completion of ω. Otherwise, move to procedure step (7).

⑦ Calculate protection angle $v = \varphi - \psi$.

⑧ Use antenna radiation reference direction figure to calculate antenna gain of ground station $G(v)$ with angle of protection v with respect to the beam.

⑨ In formula (6.52), replace G_T with gain calculated using procedure step (8), and use new r_b value to calculate the path loss L_r corresponding to hydrometeor scattering effect. If $L_r < L_b(p_2)$, increase r_b by 0.2 km and regard it as the current line-of-sight distance, otherwise, repeat from procedure step (4).

⑩ Calculate the angle θ_d with respect to position of earth station after calculating the new r_b value for current value of ω, and distance to point of contour line d is

$$d = 0.5b \sin \omega / \sin \psi_2 \tag{6.84}$$

$$\theta_d = \omega - \psi_2 \tag{6.85}$$

The auxiliary contour line of hydrometeor scattering effect is symmetrical along the main beam axis of the earth station. Therefore, corresponding values of d and θ_d when value of ω is between 181° and 359° are the same as the calculation result when $-\omega$ ~360°$-\omega$. In majority of the cases, 0.2 km increment step seems a suitable value for r_b. Consider value of r_b as fixed; this value can control the precision of the calculation result. d and θ_d can use smaller step sizes when the elevation angle of earth station's beam is small.

References

1. REC. ITU-R P.372-10 (2009) Radio noise
2. Guoliang L, Kunbi R (1990) Satellite communication. Xidian University Press, Xi'an
3. Maral G, Bousquet M (1998) Satellite communications systems-systems, techniques and technology, 3rd edn. Wiley

Chapter 7
Fade Mitigation Techniques and Propagation Analysis of Satellite System Design

The radio wave propagation effects and propagation prediction methods for satellite systems mentioned in the previous chapters are primarily utilized in satellite system link reliability design and interference coordination analysis, but can also be applied to satellite system performance evaluation. To overcome the difficulty of system interruption caused by fading, the traditional anti-fading method used is the system reserve margin method, that is, the attenuation value that should be less than the communication rate of the system is estimated via propagation, and then the power margin of the system is reserved according to the attenuation value. For satellite systems operating at Ku frequency band and above, which experience severe interruption by rain attenuation, if the high reliability requirements of the system are to be met, a sufficiently large power margin must be reserved in the uplink. Now, this not only increases the system construction and operation cost, but also appears quite challenging to achieve in areas with frequently heavy rainfall. At the same time, a very high transmit power can produce strong interference for other business systems and make the interference coordination even more taxing. For the downlink, reserving a large power margin becomes more challenging due to the load limitation of the satellite; hence, certain flexible and effective anti-jamming technologies must be adopted to ensure high reliability of the system as well as sensible use of the resources. In addition, it can also reduce the interference between different interrelated systems. This chapter mainly introduces the anti-fading technology commonly used in satellite communication systems and gives an account of the propagation design process by taking a typical satellite system as an example.

Z. Zhao et al., *Radio Wave Propagation of Satellite Systems*,
https://doi.org/10.1007/978-981-97-1394-3_7

7.1 Link Performance Analysis of Satellite Communication Systems

The basic form of satellite communication is illustrated in Fig. 7.1. The satellite link consists of two communication loops: Earth station A transmits an uplink carrier signal F_{u1} toward the satellite, while the satellite antenna and transponder receive the signal. After frequency conversion (difference frequency) and amplification, a downlink signal F_{d1} is transmitted to the earth station B. As a modulated response, the earth station B also transmits an uplink carrier signal F_{u2}, which is transmitted toward the earth station A after being forwarded by the satellite, i.e., F_{d2}.

In the satellite communication system, the transmission quality of an analog signal is determined by the ratio of the carrier power to the noise power at the input end of the receiver, that is, the (carrier-to-noise ratio, CNR) C/N ratio. When digital modulation is used, the communication quality is measured by the average bit error rate (bit error ratio, BER) (or bit error rate) P_b. The C/N ratio and bit error rate of the satellite communication system are described in the sections below.

7.1.1 Carrier-to-Noise (C/N) Ratio

1. **Satellite uplink carrier-to-noise ratio**:

Considering the basic satellite link shown in Fig. 7.1, and assuming the satellite adopts a transparent relay channel as the transponder, the Effective Isotropically

Fig. 7.1 Basic satellite communication link

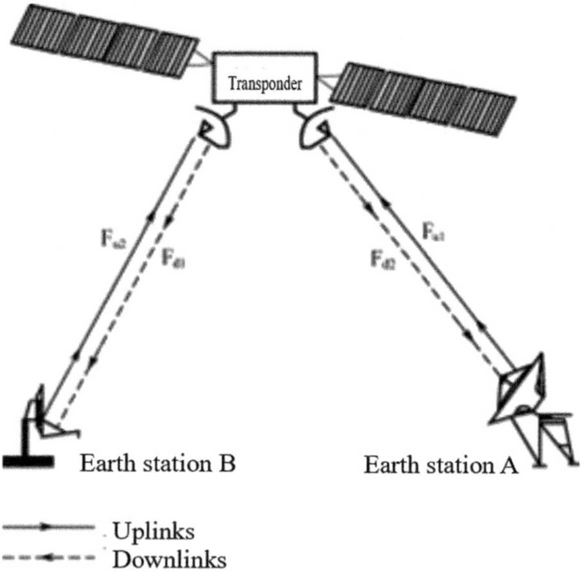

Radiated Power (EIRP) of the earth station transmitter is $P_T G_T$, where P_T is the power transmitted by the isotropic point source (W), and the transmitted antenna gain is G_T. During clear skies, the attenuation of the uplink signal only occurs in the form of free-space attenuation, gaseous attenuation, and antenna tracking loss. Under severe weather conditions, the effect of rainfall attenuation and cloud attenuation must be considered as well. The uplink free-space attenuation is

$$L_u = (4\pi d_u/\lambda_u)^2 = (4\pi d_u f_u/c)^2 \tag{7.1}$$

The form expressed in dB is as follows:

$$L_u = 92.45 + 20\lg f_u + 20\lg d_u \quad \text{dB} \tag{7.2}$$

In the formula: d_u is the uplink length (km) and f_u is the uplink carrier frequency (GHz). The summation of uplink attenuation, such as gaseous attenuation, cloud attenuation, and rain attenuation, is represented as L_u.

Assuming the gain of the satellite receiving antenna is G_u, the carrier power received by the satellite antenna is

$$C_u = (EIRP)G_u/L_u L'_u \tag{7.3}$$

The uplink noise power is given by

$$N_u = kT_u B \tag{7.4}$$

where T_u is the noise temperature (K) of the satellite receiving system, B is the noise bandwidth (Hz) of the transponder channel, and k is the Boltzmann constant (1.38×10^{-23} J/K).

Thus, the C/N ratio of the satellite uplink can be obtained as

$$\left(\frac{C}{N}\right)_u = \frac{C_u}{N_u} = \left(\frac{EIRP}{L_u L'_u}\right)\left(\frac{G_u}{T_u}\right)\left(\frac{1}{kB}\right) \tag{7.5}$$

In the formula: G_u/T_u is the gain-to-noise temperature ratio of the satellite antenna. Putting free-space attenuation into Eq. (7.5), we get

$$\left(\frac{C}{N}\right)_u = \left(\frac{EIRP}{L'}\right)\left(\frac{c}{4\pi d_u f_u}\right)^2\left(\frac{G_u}{T_u}\right)\left(\frac{1}{kB}\right) \tag{7.6}$$

If expressed in dB, it can be expressed as

$$(C/N)_u = [EIRP] + [G_u/T_u] - [L_u] - [L'_u] - [B] + 228.6 \quad \text{dB} \tag{7.7}$$

In the formula: [ ] represents 10lg (); according to the relationship $C/N_0 = C/N \times B$ (dB–Hz) between carrier wave power and noise power density ratio C/N_0, and carrier-to-noise ratio, C/N_0 can be obtained using the above calculation formula.

2. **Satellite Downlink Carrier-to-Noise Ratio and Total Carrier-to-Noise Ratio**:

The satellite receiver amplifies the received carrier-plus-noise signal, and after down-conversion or up-conversion of frequency, followed by a power amplification, it is retransmitted toward the earth by the satellite antenna. Assume that the carrier-plus-noise signal retransmitted by the satellite has the same carrier-to-noise ratio as the received carrier-plus-noise signal, that is, $(C'/N')_u = (C/N)_{\mathbf{u}}$, denote the EIRP as $\mathrm{EIRP}_{\mathbf{s}}$, and when the satellite retransmits the carrier signal, that is, $C' = \mathrm{EIRP}_{\mathbf{s}}$, so $N' = \mathrm{EIRP}_{\mathbf{s}}/(C/N)_{\mathbf{u}}$.

The propagating attenuated carrier-plus-noise voltage received by the earth station is $s'(t) + n'(t) + n_d(t)$, where $n_d(t)$ is an additional downlink added to the White Gaussian Noise with zero mean. Suppose L_d is the downlink free-space attenuation, then, L'' is the synthesis of downlink gaseous attenuation, cloud attenuation, and rainfall attenuation, and the receiving earth station antenna gain is G_d, with T_d being the noise temperature of the receiving earth station. The channel noise bandwidth is also denoted as B. Then, the total carrier signal power of the satellite downlink is

$$C = E\left[s'^2(t)\right] = \frac{EIRP_s \cdot G_d}{L_d L''} \tag{7.8}$$

The concomitant noise power in the downlink due to uplink noise is

$$N_d' = E\left[n'^2(t)\right] = \frac{N' \cdot G_d}{L_d L''} = \frac{EIRP_s}{L_d L''}\left(\frac{C}{N}\right)_u^{-1} G_d \tag{7.9}$$

The downlink noise N_d is still given by

$$N_d = E\left[n_d^2(t)\right] = kT_d B \tag{7.10}$$

Then, the total noise power at the receiving earth station is obtained as

$$N = N_d' + N_d = \frac{EIRP_s}{L_d L''}\left(\frac{C}{N}\right)_u^{-1} G_d + kT_d B \tag{7.11}$$

Therefore, the carrier-to-noise ratio of the satellite uplink and downlink can be obtained as

$$\frac{C}{N} = \frac{\left(EIRP_s \cdot G_d / L_d L''\right)}{\frac{EIRP_s}{L_d L''}\left(\frac{C}{N}\right)_u^{-1} G_d + kT_d B} = \frac{1}{\left(\frac{C}{N}\right)_u^{-1} + \left(\frac{EIRP_s \cdot G_d}{kT_d B L_d L''}\right)^{-1}} \tag{7.12}$$

where $(C/N)_u$ is the downlink carrier-to-noise ratio $\left(\frac{C}{N}\right)_d$ of the satellite.

The total carrier-to-noise ratio of the satellite uplink and downlink can be written in the following form:

$$\frac{C}{N} == \frac{1}{\left(\frac{C}{N}\right)_u^{-1} + \left(\frac{C}{N}\right)_d^{-1}} = \left[\left(\frac{C}{N}\right)_u^{-1} + \left(\frac{C}{N}\right)_d^{-1}\right]^{-1} \tag{7.13}$$

7.1.2 Bit Error Ratio

In digital signal terminology, a binary digit is called a binit. For practical applications, the information carried by a binit is defined as 1-bit information. The duration of a bit is called the bit period.

Expressed in T_b, the bit rate is $R_b = \frac{1}{T_b}$. The relationship between E_b/N_0 (the ratio of energy per bit-to-noise power spectral density) and C/N_0 (ratio of carrier-to-noise power density) is as follows:

$$E_b/N_0 = \frac{C/N_0}{R_b} \tag{7.14}$$

Taking the PSK (phase-shift keying) modulation system as an example, and assuming the noise has a flat power spectrum, its ideal bit error rate P_b can be calculated using the following formula:

$$P_b = \frac{1}{2} erfc\left(\sqrt{\frac{E_b}{N_0}}\right) \tag{7.15}$$

In the formula: $erfc(x)$ is the error complementary function, which is defined as

$$erfc(x) = \left(2/\sqrt{\pi}\right) \int_x^{\infty} e^{-u^2}\, du \tag{7.16}$$

Figure 7.1 lists the bit error rate calculation formulas of several digital modulation processes. Taking BPSK (binary phase-shift keying) as an example, its ideal bit error rate is shown in Fig. 7.2. Assuming the system requires a minimum bit error rate of 5×10^{-7}, the corresponding $[E_b/N_0]$ value is calculated to be about 10.7 dB. Assuming the clear sky $[E_b/N_0]$ value of the system is 16.5 dB, and the satellite uplink operating frequency is 30 GHz, with the downlink operating frequency being 20 GHz, the satellite fixed point longitude is around 100.5°, and the ground station is in the Qingdao area. The rain attenuation of 99.5% availability of the uplink predicted by the rain attenuation prediction model is 11 dB, that is, the uplink $[E_b/N_0] = 16.5 - 11 = 5.5$ dB, then 5.2 dB of energy needs to be compensated to achieve the minimum bit error rate standard. The predicted downlink rain attenuation with the same probability is 4.9 dB, and the rain noise is given by

$$T_s = 260(1 - 10^{-4.9/10}) \approx 175.9 \quad \text{K} \tag{7.17}$$

If the effective noise temperature of the receiving system is 400 K, the increase in noise power is

$$10\lg\frac{400 + 175.9}{400} \approx 1.6 \quad \text{dB} \tag{7.18}$$

Hence, the downlink $[E/N_0] = 16.5 - 4.9 - 1.6 = 10$ dB, which needs to be compensated by 0.7 dB (Table 7.1).

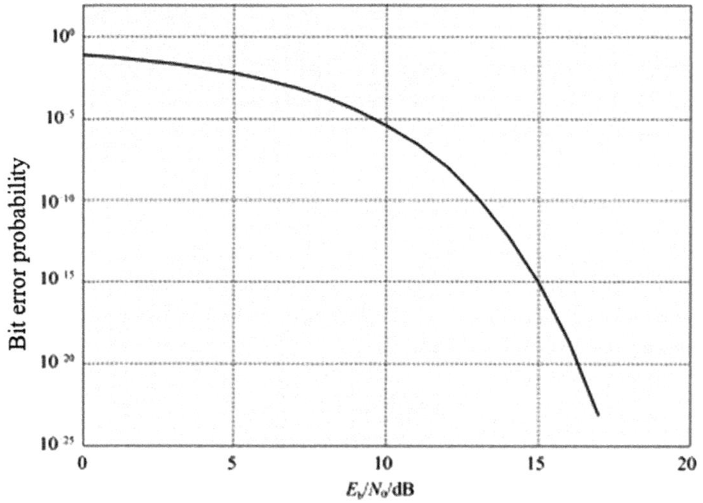

Fig. 7.2 Ideal BPSK modulation performance

Table 7.1 Bit error ratio of typical modulation methods [1]

System modulation type		Bit error ratio
Direct encoding	BPSK	$1/2erfc\left(\sqrt{E_b/N_0}\right)$
	QPSK	$1/2erfc\left(\sqrt{E_b/N_0}\right)$
Differential encoding	DE-BPSK	$erfc\left(\sqrt{E_b/N_0}\right)$
	DE-QPSK	$erfc\left(\sqrt{E_b/N_0}\right)$
	D-BPSK	$1/2\exp(-E_b/N_0)$

7.2 Fade Mitigation Techniques for Satellite Communication Systems

For satellite communication systems operating in the Ku frequency band and above, the satellite-ground link attenuation tends to get increasingly acute, and the traditional method of reserve margin fails to meet the reliability requirements of the system. Anti-fading technology is required to cut back on the abrupt attenuation of the link (mainly rain attenuation) and to compensate it suitably. Prevalent anti-fading technologies can be divided into three categories, namely, diversity technology, power control technology, and adaptive signal processing technology, as shown in Fig. 7.3. Practically, multiple anti-fading technologies are often put into operation at the same time. For example, the ACTS satellite communication system in the United States comprehensively implements link margin, uplink power control technology, adaptive error correction coding technology, and adaptive rate adjustment technology. This section introduces and evaluates the anti-fading techniques commonly applied in satellite communication systems.

7.2.1 Diversity

Diversity in telecommunication refers to the utilization of multiple communication channels to combat link propagation attenuation and enhance the reliability of a signal. There exist several diversity methods, such as site diversity, orbit diversity, and frequency diversity.

1. Site diversity:

Due to the inhomogeneity of rainfall distribution, particularly heavy rainfall events, such as torrential rain, always occur within a certain scale, (site diversity SD) which is essentially based on this principle. For better reception of satellite signals under reduced influence of rain attenuation, ideally, the two sites should have independent rainfall distribution characteristics, and the distance between the sites must at least

- *Rain attenuation mitigation techniques*
  - *Diversity technique*
    - *Site diversity*
    - *Orbit diversity*
    - *Frequency diversity*
  - *Power control technique*
    - *Uplink power control technique*
      - *Open loop*
      - *Closed loop*
      - *Feedback loop*
    - *Downlink power control technique*
  - *Self – adaption signal processing technique*
    - *Self – adaption correction coding*
    - *Self – adaption modulation*
    - *Self – adaption transmission speed regulation*

Fig. 7.3 Classification of commonly used fade mitigation techniques in satellite communication systems

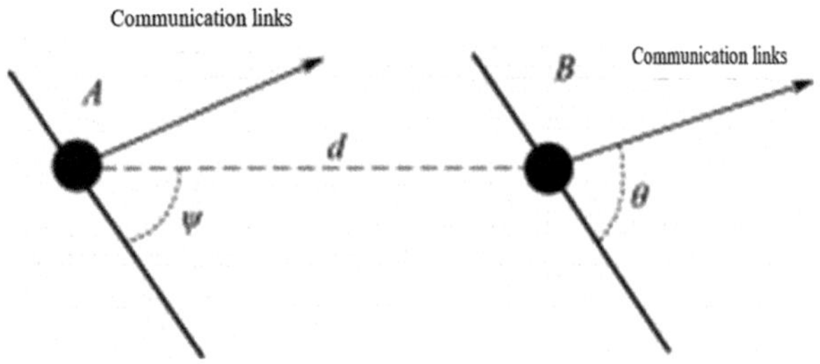

Fig. 7.4 Schematic diagram of site diversity

exceed the rain cell or rain belt radius during a heavy rainfall event. It is generally believed that the distance above 10 km can fulfill these requirements.

Figure 7.4 is a schematic diagram of site diversity, where θ is the elevation angle of the satellite-ground path; d is the baseline distance between two sites A and B; and ψ is the angle between the baseline and the satellite link projected on the ground, called the baseline azimuth angle. Baseline distance and baseline azimuth angle emerge as the two most critical parameters affecting the diversity effect. As the baseline distance increases, the instantaneous correlation of rain attenuation between the two sites decreases; the ideal angle of the baseline azimuth is 90°, at which time the two satellite links are optimally isolated in space. Site diversity is primarily suitable for fixed satellite services. The experimental results portray that the method can increase the link gain by 10–30 dB for the Ka-band satellite system.

The functioning of site diversity is often examined by the joint probability or diversity gain of rain attenuation exceeding a given threshold. The following two diversity stations are used as examples to introduce the corresponding prediction methods [2].

(1) Prediction method for rain attenuation joint probability distribution of site diversity.

This diversity prediction method considers that rainfall and rain attenuation events follow a lognormal distribution. Let $P_r\ (A_1 \geq a_1, A_2 \geq a_2)$ denote the joint probability distribution of the rain attenuation for the two concerned diversity stations, that is, the occurrence probability of the rain attenuation of the first station of diversity $A_1 \geq a_1$ and the rain attenuation of the second station of $A_2 \geq a_2$. Defining p_{r0} as the probability of simultaneous rainfall at the two stations, and pa as the conditional joint probability that shows the rain attenuation exceeds a_1 and a_2, respectively, in the case of rainfall at both the stations, that is, $p_a = p\ (A_1 \geq a_1, A_2 \geq a_2 \mid p_{r0})$. Here [3]:

$$p_r(A_1 \geq a_1, A_2 \geq a_2) = 100 \times p_{r0} \times p_a(\%) \tag{7.19}$$

The respective calculation formulas of p_{r0} and p_a are

$$p_{r0} = \frac{1}{2\pi\sqrt{1-\rho_r^2}} \int_{R_1}^{\infty}\int_{R_2}^{\infty} \exp\left[-\left(\frac{r_1^2 - 2\rho_r r_1 r_2 + r_2^2}{2(1-\rho_r^2)}\right)\right] dr_1 dr_2 \tag{7.20}$$

$$p_a = \frac{1}{2\pi\sqrt{1-\rho_a^2}} \int_{\frac{\ln a_1 - m_{\ln A_1}}{\sigma_{\ln A_1}}}^{\infty} \int_{\frac{\ln a_2 - m_{\ln A_2}}{\sigma_{\ln A_2}}}^{\infty} \exp\left[-\left(\frac{a_1^2 - 2\rho_a a_1 a_2 + a_2^2}{2(1-\rho_a^2)}\right)\right] da_1 da_2 \tag{7.21}$$

ρ_r and ρ_a can be defined as

$$\rho_r = 0.7\exp(-d/60) + 0.3\exp\left[-(d/700)^2\right]$$
$$\rho_a = 0.94\exp(-d/30) + 0.06\exp\left[-(d/500)^2\right]$$

Thresholds R_1, R_2 are solutions of the following equations:

$$p_k^{rain} = 100 \times Q(R_k) = 100 \times \frac{1}{\sqrt{2\pi}} \int_{R_k}^{\infty} \exp\left(-\frac{r^2}{2}\right) dr \tag{7.22}$$

where

$$R_k = Q^{-1}\left(\frac{p_k^{rain}}{100}\right) \tag{7.23}$$

In the formula: $m_{\ln A1}$, $m_{\ln A2}$, $\sigma_{\ln A1}$, and $\sigma_{\ln A2}$ are the lognormal distribution parameters of the station rain attenuation, which can be obtained by lognormal fitting between the station rain attenuation A_i and the corresponding time probability p_i. The fitting formula is shown in formula (7.24), and the specific fitting process can be found in Sect. 3.4.1 of Chap. 3.

$$p_i = p_k^{rain} Q\left(\frac{\ln A_i - m_{\ln A_i}}{\sigma_{\ln A_i}}\right) \tag{7.24}$$

In the formula: R_k is the threshold of the kth station ($k = 1, 2$) and p_k^{rain} is the rain probability of the kth station ($k = 1, 2$).

Taking the satellite-ground link of a certain site as an example, the operating frequency is 20 GHz, the link elevation angle is 43.4°, the probability of rain at the site is 3.9%, and the rainfall rate that exceeds 0.01% of the time is 78.27 mm/h. It can be calculated that the 0.1% time probability of a single station is exceeded by the rain attenuation of 11.72 dB, this means the system needs to compensate for the attenuation of nearly 12 dB to achieve a passability rate of 99.9%. If the site diversity technology is utilized, and the diversity distance is 20 km, the above prediction method can be used to obtain the diversity joint probability p_r ($A_1 \geq 12$, $A_2 \geq 12$), which is approximately 0.01%, which indicates that compensating for the same

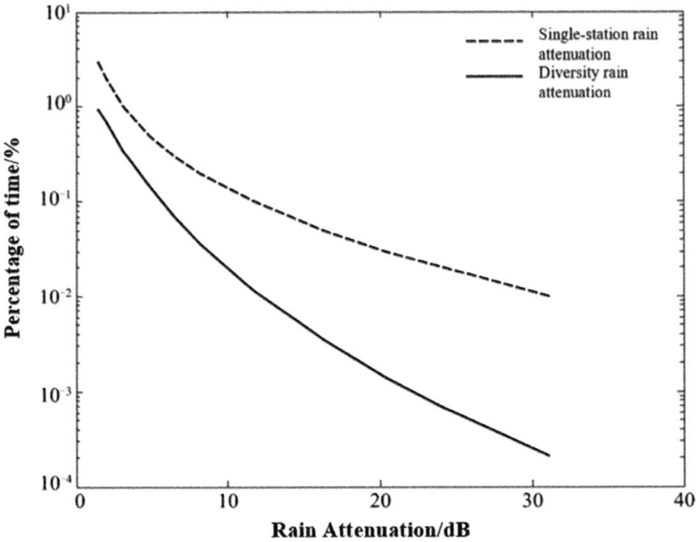

Fig. 7.5 Cumulative distribution of rain attenuation for single station and site diversity

rain attenuation after implementing diversity can increase the passability rate of the system up to 99.99%. Figure 7.5 shows the cumulative distribution of rain attenuation in case of single station and site diversity of the above link. It can be observed that the site diversity can effectively circumvent the influence of rain attenuation on the system, and thereby enhance the passability of the system significantly.

(2) Diversity Gain Prediction Method:

This method is suitable for the case where the distance of the diversity site is less than 20 km and the frequency ranges between 10 GHz $\leq f \leq$ 30 GHz. The prediction formula for site diversity gain is as follows:

$$G_{sd} = G_d \cdot G_f \cdot G_\theta \cdot G_\psi \tag{7.25}$$

where G_d is the distance gain factor between the two stations, which can be defined as

$$G_d = a\left(1 - e^{bd}\right) \tag{7.26}$$

$$a = 0.78A - 1.94(1 - e^{-0.11A})$$
$$b = 0.59(1 - e^{-0.1A})$$

G_f is the frequency gain factor, which can be defined as

$$G_f = e^{-0.025f} \tag{7.27}$$

G_θ is the elevation angle gain factor, which can be defined as

$$G_\theta = 1 + 0.006\theta \tag{7.28}$$

G_ψ is the baseline azimuth gain factor, which can be defined as

$$G_\psi = 1 + 0.002\psi \tag{7.29}$$

Considering a satellite-to-ground link with a frequency of 20 GHz as an example, the link elevation angle is 50.4°, and the 0.1% ($p_1 = 0.1\%$) time probability of being exceeded by rain attenuation is 11.18 dB. The site diversity parameters are as follows: $d = 10$ km, $\psi = 85°$. Then, the diversity gain is calculated as follows:

$$\begin{gathered} a = 0.78 \times 11.18 - 1.94 \times \left(1 - e^{-0.11 \times 11.18}\right) = 7.35; \\ b = 0.59 \times \left(1 - e^{-0.1 \times 11.18}\right) = 0.40; \\ G_d = a\left(1 - e^{-bd}\right) = 7.22; \\ G_f = 0.61,\ G_\theta = 1.12,\ G_\psi = 1.17; \\ G = G_d \cdot G_f \cdot G_\theta \cdot G_\psi = 5.77 \quad (\text{dB}). \end{gathered}$$

It is evident that the site diversity brings a gain of 5.77 dB. Figure 7.6 illustrates the effect of baseline distance d and baseline azimuth ψ on diversity gain. After the baseline distance exceeds 10–20 km, the diversity gain barely shows any variation. The research attributes this to the experimental data adopted for the modeling. This interpretation needs further validation by additional experiments.

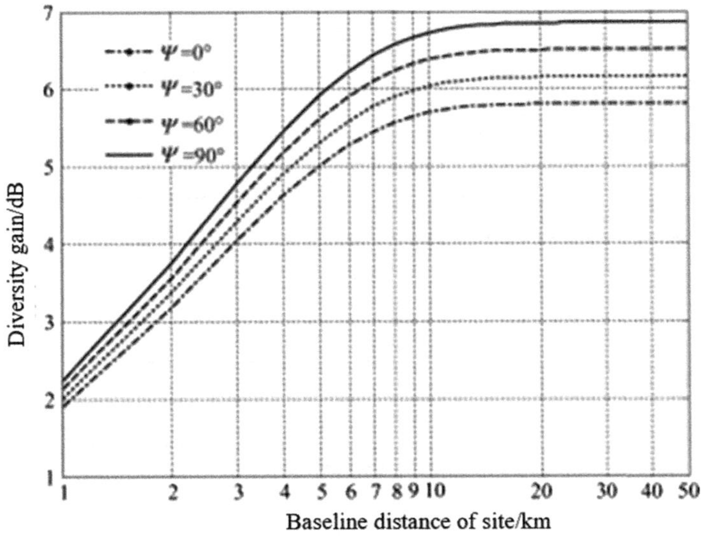

Fig. 7.6 Relationship between diversity gain, baseline distance, and azimuth

2. Orbit Diversity and Frequency Diversity:

(1) Orbit Diversity:

Orbit Diversity (OD) is divided into two ways: One is to optimize the number and arrangement of satellites to avoid communication lapse in a low-elevation environment. The other is to allow the ground station to choose a more reasonable link for communication between the two satellites with different orbital positions, according to the path propagation environment. The second case is mainly discussed here. Since the ground station is fixed, the diversity paths of orbit diversity possesses strong correlation; hence, the anti-fading effect of orbit diversity does not appear as evident as that of site diversity. The advantage of orbit diversity's resource sharing mechanism over site diversity's anti-fading technology is that owing to a reasonable means of resource sharing, satellites can offer resource sharing for multiple ground stations, which is an effective way of enhancing the resource utilization efficacy. In addition, orbit diversity also appears conducive in the event of failure of the preferred satellite or other related emergencies.

Using the COMSTARD1 and COMSTARD2 satellites, the S1R10 and OTS satellites, and the OLYPUS and ITALSAT satellites, the researchers conducted experimental studies on orbit diversity, and realized that the orbit diversity gain decreases with the increase of the link elevation angle, and the diversity gain effect becomes more detectable with the increase in frequency, and the normalized orbit diversity gain can be increased by 30% at 12 GHz.

The assumption of orbit diversity is that there are satellites to choose from. At the same time, to save the switching time between different links, the ground station generally needs to be equipped with at least two sets of antenna systems.

(2) Frequency Diversity:

Frequency Diversity (FD) technology requires the satellite to carry loads of various frequency bands. When there occurs an acute attenuation event in the Ka frequency band and above, such as rainfall, the lower X frequency band or Ku frequency band is selected for communication. Applying frequency diversity requires the ground station to have two sets of receivers. Experiments demonstrate that when the system switches between the Ku frequency band and the Ka frequency band, a gain of up to 30 dB can be obtained.

7.2.2 Power Control

The power control technology suitably changes the transmit power according to the real-time attenuation of the satellite-ground link, to ensure that the signal power received by the receiver fulfills the communication requirements. The power control technology is also applicable to the uplink and the downlink, which are called Uplink Power Control (ULPC) and Downlink Power Control (DLPC), respectively. Power control technology holds remarkable advantages over mere compensation for the link attenuation with high transmit power. On one hand, it can prevent the satellite transponder from being overloaded, and on the other hand, it can avert the interference of long-term high-power transmission with other links and hence save resources. The maximum link attenuation that the power control technique can compensate for is equal to the difference between the maximum output power of the ground station or satellite power amplifier and the output power required for the link to establish a proper communication channel under clear sky conditions.

1. **Uplink Power Control Technology** [4]:

Uplink power control technology is the most commonly utilized satellite anti-fading technology. When tropospheric events, such as rainfall, occur, the transmit power of the ground end is increased to compensate for the attenuation on the link; in the case of a clear sky, the transmit power of the ground end is reduced to avoid wastage of resources. The ultimate objective is to ensure that the signal strength reaching the satellite transponder remains constant.

There are three ways of operating uplink power control—pen loop, closed loop, and feedback loop. Although closed-loop and feedback-loop control can significantly enhance the control accuracy, a certain delay is observed (closed loop 0.26 s, feedback loop 0.52 s), and the system is more intricate than the open-loop method. For practical applications, open-loop control emerges more convenient to implement and has been the most widely studied method. By suitably adjusting the parameters of open-loop control, better control accuracy can also be achieved. In the open-loop method, the satellite link attenuation is estimated by the ground-side observation equipment together with the propagation prediction model, and the transmit power of the ground station is adjusted accordingly. In general, the ground terminal monitors the link attenuation (mainly rain attenuation) through the satellite downlink beacon signal or the gaseous radiation signal received by the microwave radiometer.

When monitoring the link attenuation based on the satellite downlink beacon signal, the uplink attenuation needs to be obtained through the conversion model. For the specific method, see Sect. 3.4 in Chap. 3. When employing the microwave radiometer to monitor the link attenuation, the calculation formula of the link attenuation is as follows:

$$A = 10\lg \frac{T_m}{T_m - T_s} \tag{7.30}$$

where A is the link attenuation (dB), T_m is the effective atmospheric radiation temperature, generally 260 K, and T_s is the link radiation brightness temperature (K) observed by the microwave radiometer. The microwave radiometer can get saturated, and experiments and studies have shown that the link attenuation that can be monitored by the radiometer should not exceed 13 dB.

2. **Downlink Power Control**

The downlink power control technology ensures that the signal reaching the ground terminal can fulfill the communication requirements by increasing the satellite downlink transmission power. The downlink power control technology generally compensates the downlink attenuation in two ways: one is to increase the power of the satellite transmitter and the other is to adjust the satellite transmit antenna gain according to the propagation environment to achieve the purpose of changing the satellite EIRP, also known as the satellite carrier beam formation (on-board beam shaping, OBBS) technology. The implementation of downlink power control technique is far more complex than that of the uplink power control technique.

When downlink power control technology is applied, the received signal power in the area covered by the satellite beam gets enhanced, which causes interference to adjacent ground stations. Constrained by the satellite payload, the link attenuation that can be compensated by increasing the power of the satellite transmitter is limited. The beam forming technology is based on the space-borne active antenna array, and the purpose of increasing the antenna gain, i.e., increasing the effective transmission power, is achieved by reducing the size of the spot beam. Therefore, the uplink power control technology is the most prevalent one for practical applications.

7.2.3 *Adaptive Signal Processing*

Adaptive signal processing (ASP) technology chiefly includes adaptive error correction coding, adaptive modulation, and adaptive transmission rate adjustment technology.

1. Adaptive Error Correction Coding:

The adaptive error correction coding technology compensates for the propagation attenuation of the link by changing the coding rate. Considering the time-division multiple access system as an example, Adaptive Forward Error Correction (AFEC) is usually adopted to reduce the bit error rate, which can generally compensate for the attenuation of 2–10 dB. A time-division multiple access satellite communication system with 32 ground terminals operating at 11/14 GHz can achieve a coding gain of up to 8 dB using AFEC technology. The American ACTS communication satellite also utilizes the AFEC technology, which can compensate for the 10 dB attenuation with the rate reduction technology.

2. Adaptive modulation:

Adaptive modulation technology is an adaptive transmission rate technology, where the information bit rate changes with the modulation method. As an anti-fading technology, adaptive modulation technology avoids the utilization of a large fixed fading reserve to obtain high availability, reducing resource wastage and facilitating the system to reach the maximum limit of its capacity, along with guaranteeing high communication quality and data transmission volume. Through the statistical analysis of Bit Error Rate (BER) of the system, it is found that under ideal conditions, the interruption time of the adaptive modulation system is ¼ times less than that of the traditional 8-PSK system, and the data output increases by about 30% in 1 year.

However, adaptive modulation also presents new requirements for the system: Firstly, the system requires proper communication equipment to support multiple modulation schemes, and, secondly, it is vital to recognize the intelligence possessed by the system for self-adaptation. In addition, the realization of an adaptive modulation system must be premised on the reduction of the instantaneous data rate that the system can tolerate.

3. Adaptive transmission rate adjustment:

The adaptive transmission rate adjustment is to adaptably adjust the transmission rate of the signal according to the propagation attenuation of the satellite-ground link. The gain obtained by using this technology is proportional to the ratio of the standard signaling data rate to the reduced signaling data rate. The adaptive transmission rate adjustment technology needs the system to reserve frame time slots to achieve burst rate reduction, and the relative effective available capacity of the system is

$$C_{ARR} = \frac{1}{1 + F(R_D - 1)} \tag{7.31}$$

In the formula: F is the ratio of the number of channels that need to be reduced to the total number of channels; R_D is the ratio of the standard signal data rate to the reduced signal data rate, which is equal to the time frame extension. When the transmission rate is reduced from 2048 kb/s to 1024, 512 to 256 kb/s, in turn, gains of 3, 6, and 9 dB can be obtained, respectively.

7.3 Propagation Analysis of Typical Satellite System Design

The satellite systems operate at bands ranging from UHF to EHF, and radio waves in different bands are affected differently by the ionosphere and troposphere. For example, satellite systems using low-frequency bands are primarily affected by ionospheric propagation effects, while the ones working at high-frequency bands (greater than 3 GHz) are mainly affected by convection layer. These propagation effects must be considered in system design. For the 3–10 GHz band, due to the small influence of ionospheric propagation effect, mainly tropospheric gaseous attenuation, cloud attenuation and tropospheric scintillation, rain attenuation, and rain-induced depolarization effect can be considered. The design process can refer to the satellite system design above 10 GHz band. In this chapter, the propagation effects and broadcast design process of satellite systems operating below 3 GHz and above 10 GHz are introduced, along with a brief introduction about the propagation design of mobile-satellite systems.

7.3.1 *Propagation Analysis of Fixed Satellite System Design*

7.3.1.1 Propagation Design of a Satellite System Operating in Frequency Band Below 3 GHz [5]

For satellite systems operating in the frequency band below 3 GHz, the propagation effect generated by the radio wave environment mainly occurs in the ionosphere, and the effects of Faraday rotation, group delay, and time delay dispersion caused by the ionosphere need to be considered. Furthermore, in some cases, allowances must be also made for ionosphere, scintillation, polar zone absorption, and polar-cap absorption. The associated content is described thoroughly in Chap. 5 of this book.

Considering the impact of the ionosphere on the satellite system, the following parameters need to be determined: system frequency f (GHz); percentage time to reach preset performance indicators (annual deviation); polarization deflection angle T (°); link elevation angle θ (°); the latitude φ (°) of the ground station, represented by north latitude N or south latitude S, the north latitude is positive and the south latitude is negative; the longitude δ (°) of the ground station, represented by east longitude E or west longitude W, east longitude is positive, and the west longitude is negative; and the altitude is h_s (km).

Majority of the ionospheric propagation effects require the determination of the total electron content of the ionospheric path, TEC; thus, the accurate calculation of TEC for the signals in the ionosphere is the first crucial step. For some systems, if accurate TEC is not available, the TEC value can be estimated using the following method: TEC $= N_m F_2 \times H_{\text{slab}}$, where $N_m F_2$ represents the maximum electron density of the F_2-layer and H_{slab} represents the equivalent plate thickness of the ionosphere, which is generally 300 km.

Figure 7.7 demonstrates the propagation design process of the satellite system in the frequency band below 3 GHz. The calculation method of ionosphere-related effects is described in depth in Chap. 5.

7.3.1.2 Propagation Design of Satellite Systems Above 10 GHz [5]

Satellite systems operating at Ku/Ka/Q/V bands are predominantly affected by tropospheric propagation effects. This section introduces the propagation effects that are crucial to be considered in the design of geostationary satellite systems for the above-mentioned frequency bands.

1. Propagation design of Ku-band satellite system:

This section mainly introduces the propagation design process of Ku-band Fixed Satellite Service (FSS) system, low-margin Ku-band FSS system, and Broadcast Satellite Service (BSS) system.

(1) Propagation design of Ku-band FSS system [5]:

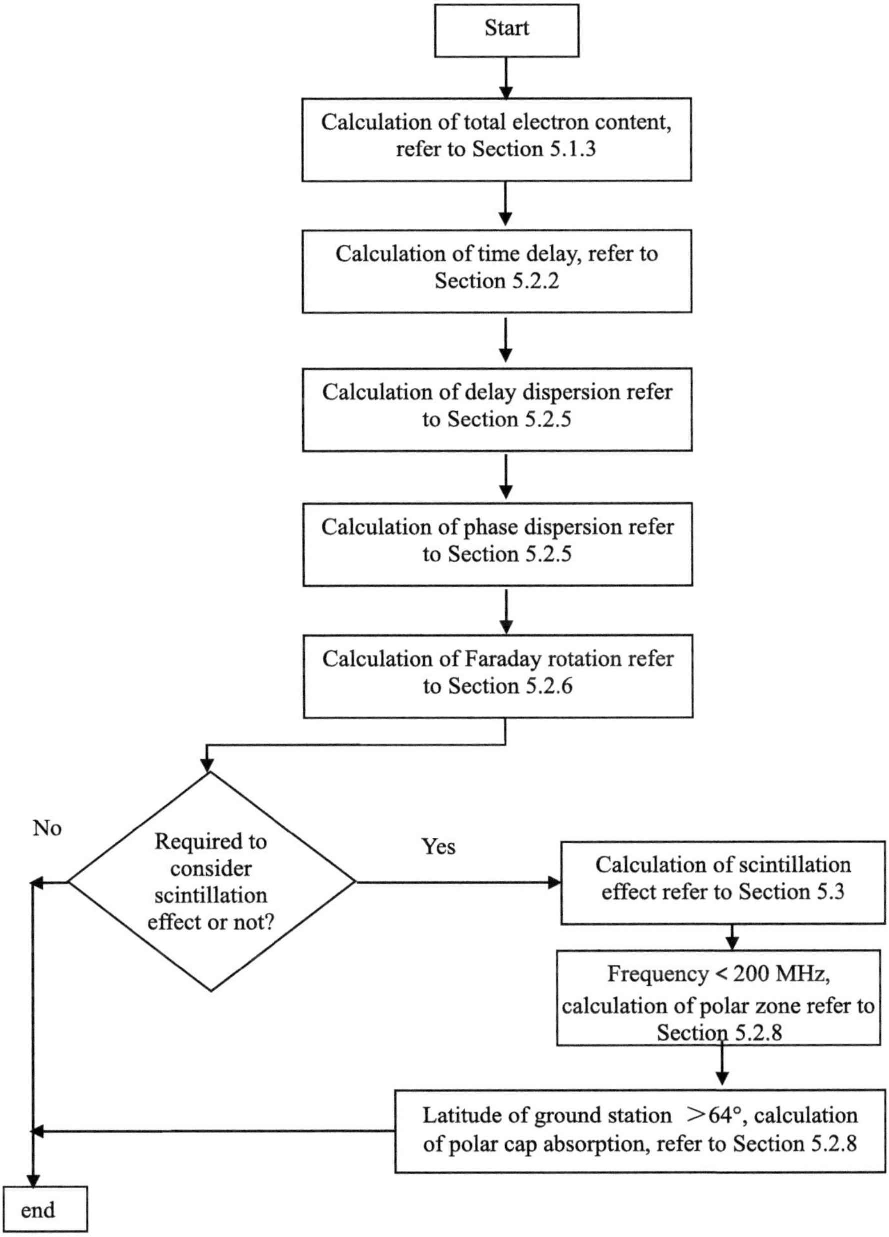

Fig. 7.7 Flowchart of satellite system propagation design in frequency bands below 3 GHz

The propagation effect prediction method is applied to the Ku-band FSS system design, and the final system parameters (antenna size, beam width, transmission power, etc.) are determined based on the overall performance requirements of the system (link availability, interruption time, etc.). The passability of the Ku-band FSS system design is generally 95–99.99%, and the influence of rain attenuation is primarily considered in the system design. If the elevation angle of the space-earth link is lower than 20°, tropospheric scintillation fading must be considered as well. After measuring the above propagation effects by using diverse propagation effect prediction methods provided in Chaps. 2 and 3, comprehensive attenuation can be calculated by using the method of gaseous comprehensive attenuation provided in Sect. 3.10. When the system adopts frequency multiplexing, the rain-induced depolarization effect and ice crystal depolarization effect need to be considered as well. Figure 7.6 illustrates the propagation design process of the Ku frequency band FSS system. Please refer to the relevant previous sections in this book for calculation methods of different propagation effects (Fig. 7.8).

When the system utilizes anti-fading technology, the further design flowchart is shown in Fig. 7.9.

(2) Propagation design of low-margin Ku-band FSS system [5]:

Low-margin Ku-band FSS systems generally adopt very small-aperture terminal (VSAT) systems, and their power margin is quite low, usually 1–3 dB, with the corresponding system pass rate being 99–99.5%. To fulfill the requirements of the possibility, together with rain attenuation, propagation effects, such as cloud and gaseous attenuation, need to be considered as well. The propagation design flowchart of the low-margin Ku-band FSS system is shown in Fig. 7.10.

(3) BSS system propagation design:

Currently, the downlink of the direct-to-household BSS system predominantly works in the Ku frequency band range, and the specific frequency band allocation is 11.7–12.5 GHz in the first area, 12.2–12.7 GHz in the second area, and 11.7–12.2 GHz in the third area. The uplink generally works at 17.3–16.1 GHz. The propagation design of the BSS system must make allowances for effects of gaseous attenuation, rain attenuation, and cloud attenuation. Ground terminals generally use small-aperture antennas, and the EIRP of the satellite system is 6 to 10 dB larger than that of the FSS system. Its system performance is not evaluated using annual average statistics but is determined by the worst-month link availability. Figure 7.11 shows the propagation design flowchart of the BSS system.

2. Propagation design of Ka-band satellite system [5]:

At present, the Ka-band satellite system is widely utilized. This section introduces the propagation design of the Ka-band FSS system and the low-margin Ka-band FSS system.

(1) Ku-band FSS system:

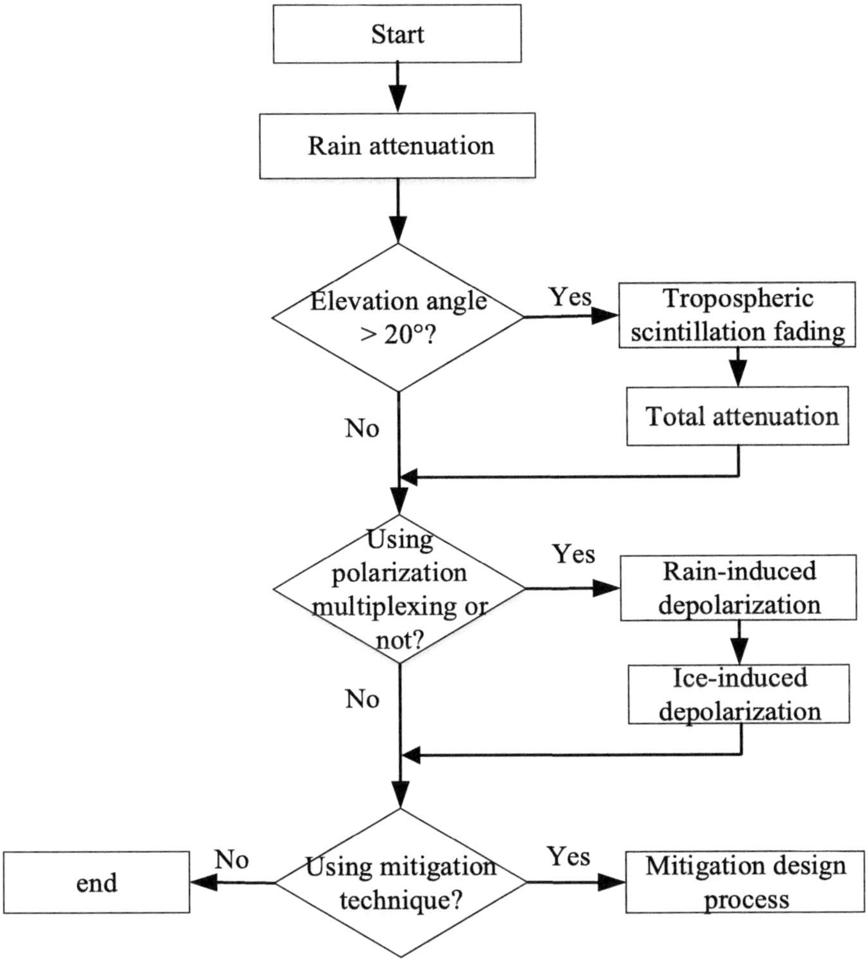

Fig. 7.8 Ku-band FSS system propagation design flowchart

In addition to gaseous attenuation and rain attenuation, the propagation design of Ka-band FSS system must consider the tropospheric scintillation effect as well. When the system adopts frequency multiplexing, it is also essential to consider the rain-induced depolarization effect and the ice crystal depolarization effect. The Ka-band FSS system propagation design flowchart is shown in Fig. 7.12, and the anti-fading technology design flowchart is shown in Fig. 7.9.

(2) Low-margin Ka-band FSS system:

Analogous to the low-margin Ku-band FSS system, the low-margin Ka-band FSS system generally adopts the VSAT system, where the link power margin is generally 1–3 dB, and the corresponding system passability is 99–99.5%. The influence of

Fig. 7.9 Flowchart for fade mitigation technique design

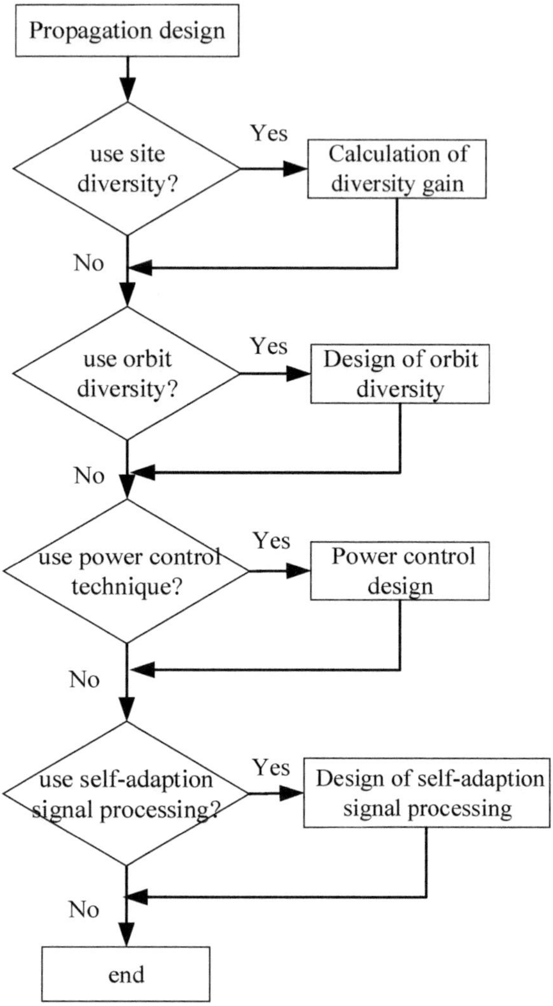

gaseous attenuation, cloud attenuation, and rain attenuation needs to be considered for the system design. Figure 7.12 demonstrates the propagation design flowchart of the low-margin Ka-band FSS system, and the design flow of anti-fading technology is shown in Fig. 7.9.

3. Propagation design of Q/V frequency band satellite system:

Q/V frequency band FSS system's uplink frequency range is 47.2–50.2 GHz, the downlink frequency range is 37.5–40.5 GHz, data transmission rate is up to 3 Gb/s and is mainly utilized for broadband multimedia services, VSAT, and direct-to-household services. Propagation effects that need to be considered in the design of Q/V band FSS systems include gaseous attenuation, rain attenuation, and cloud

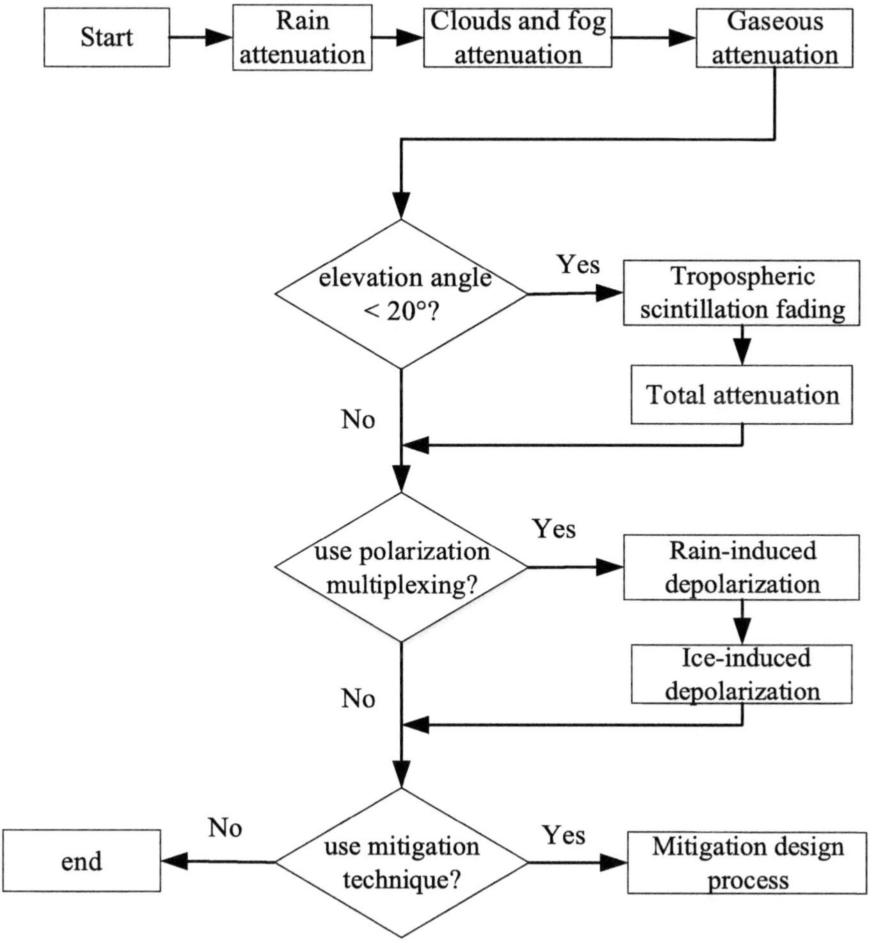

Fig. 7.10 Low-margin Ku-band FSS system propagation design flowchart

attenuation. When the elevation angle of the satellite-ground link is less than 20°, the tropospheric scintillation effect needs to be considered as well. When the system adopts frequency reuse or multiplexing, it also becomes essential to consider the rain-induced depolarization effect and ice crystal depolarization effect. Figure 7.13 shows the propagation design flowchart of the Q/V frequency band FSS system, and the design flowchart of the anti-fading technology is shown in Fig. 7.9.

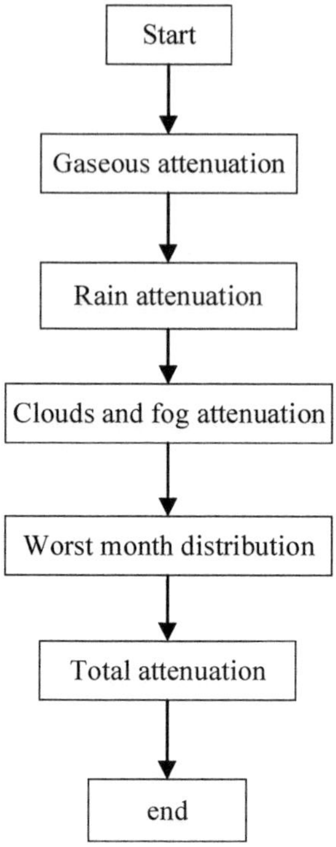

Fig. 7.11 BSS system propagation design flowchart

7.3.2 Propagation Analysis of Mobile-Satellite Service System Design

In addition to the aforementioned effects produced by the troposphere and ionosphere on the satellite systems, the mobile-satellite system also needs to weigh the various unique propagation effects stimulated under diverse application scenarios in the system design. For example, the maritime mobile-satellite system mainly considers the reflection of the sea surface and the interference effect of adjacent satellites; the land mobile-satellite system primarily considers the shading effect of land, vegetation, and buildings, as well as the statistical calculation method under mixed propagation conditions; and the aeronautical mobile-satellite system mostly only considers the sea surface. The reflection fading effect and the propagation effect brought about by the multipath calculation method of approach and landing of an aircraft are shown in Fig. 7.14. Refer Chap. 4 for a detailed introduction as well as the particulars of the corresponding propagation calculation methods.

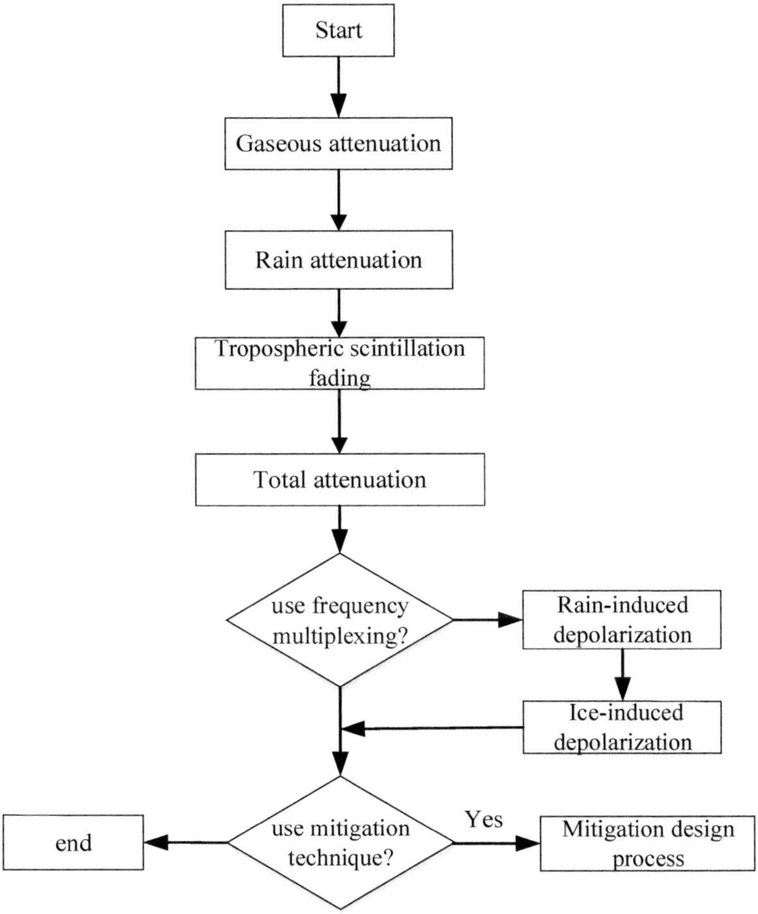

Fig. 7.12 Flowchart of propagation design of Ka-band FSS system

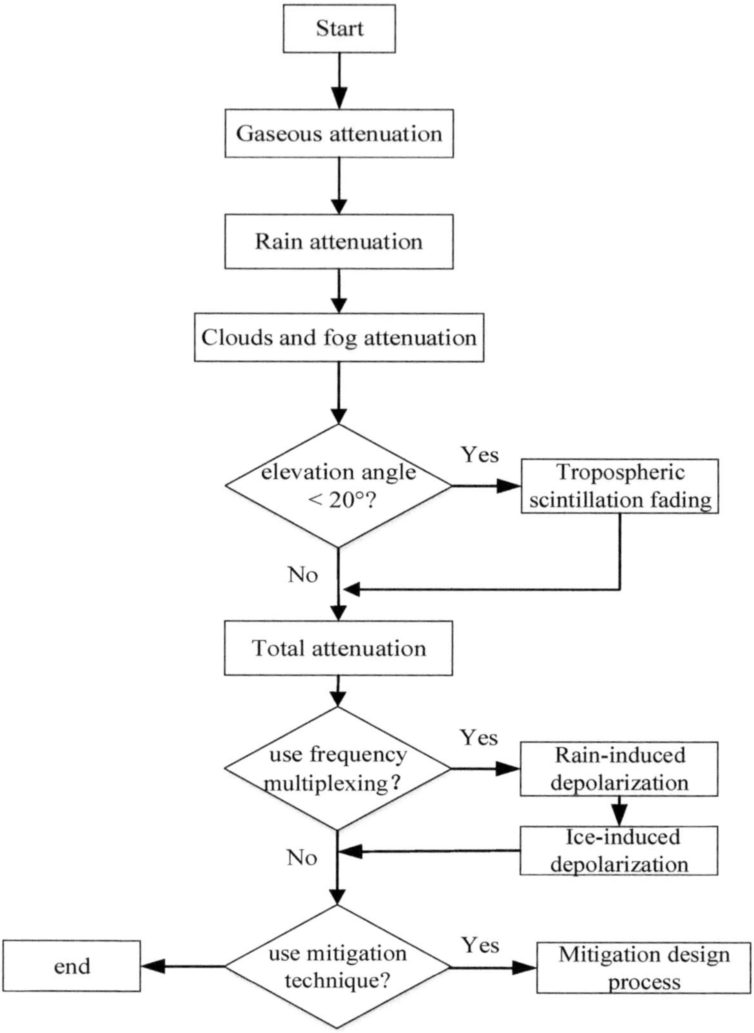

Fig. 7.13 Flowchart of the propagation design of the Q/V frequency band FSS system

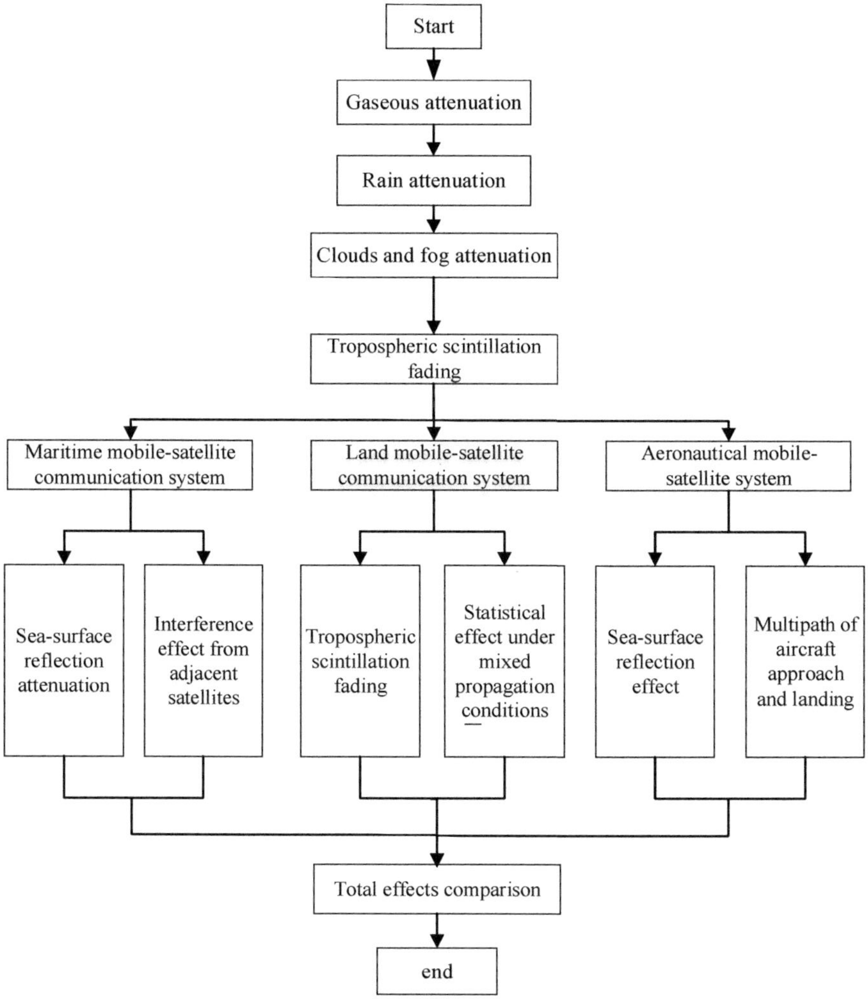

Fig. 7.14 Propagation design of mobile-satellite service systems

References

1. Roddy D, Gengxin Z et al (2002) Satellite communications, 3rd edn. Posts and Telecommunications Press, Beijing
2. Zhenguo C, Hongwen Y, Wenbin G (2003) Satellite communication system and technology. Beijing University of Posts and Telecommunications Press, Beijing
3. International Telecommunications Union (ITU) (2002) Handbook on satellite communications, 3rd edn. Wiley-Interscience
4. Ippolito LJ (1999) Propagation effects handbook for satellite systems design, 5th edn. Section 3, Applications. Jet Propulsion Laboratory

5. Ippolito Jr LJ (2008) Satellite communications systems engineering: atmospheric effects, satellite link design and system performance. Wiley

GPSR Compliance

The European Union's (EU) General Product Safety Regulation (GPSR) is a set of rules that requires consumer products to be safe and our obligations to ensure this.

If you have any concerns about our products, you can contact us on ProductSafety@springernature.com

In case Publisher is established outside the EU, the EU authorized representative is:

Springer Nature Customer Service Center GmbH
Europaplatz 3
69115 Heidelberg, Germany

Batch number: 10732090

Printed by Printforce, the Netherlands